Molecular Biology Intelligence Unit

The Biology of Genetic Dominance

Reiner A. Veitia, Ph.D.
Université Denis Diderot/Paris VII
Paris, France

CRC Press
Taylor & Francis Group
Boca Raton London New York

CRC Press is an imprint of the
Taylor & Francis Group, an **informa** business

The Biology of Genetic Dominance

Molecular Biology Intelligence Unit

CRC Press
Taylor & Francis Group
6000 Broken Sound Parkway NW, Suite 300
Boca Raton, FL 33487-2742

First issued in hardback 2019

ISBN-13: 978-1-58706-288-9 (hbk)

Visit the Taylor & Francis Web site at
http://www.taylorandfrancis.com

and the CRC Press Web site at
http://www.crcpress.com

Library of Congress Cataloging-in-Publication Data

The biology of genetic dominance / edited by Reiner A. Veitia.
p. ; cm. -- (Molecular biology intelligence unit)
Includes bibliographical references and index.
ISBN 1-58706-288-7
1. Gene expression. 2. Phenotype. I. Veitia, Reiner A. II. Series: Molecular biology intelligence unit (Unnumbered) [DNLM: 1. Gene Expression--genetics. 2. Genes, Dominant--physiology. 3. Phenotype. QH 450 B6158 2005]
QH450.B55 2005
611'.01816--dc22

2005005137

CONTENTS

Preface ix

1. **The Basis of Dominance** 1
Athel Cornish-Bowden and Vidyanand Nanjundiah
Questions 2
Framework within Which to Seek Possible Answers 3
Dominance As a Product of Evolution 3
Wright's Theory: Dominance As a Physiological Phenomenon 5
Effect of Enzyme Activity on Metabolic Flux 5
Effect of High Saturation of Enzymes 9
Dominant Genes in Humans: An Exceptional Species? 11
The Hexokinase D ("Glucokinase") Gene in Humans 11
Another Look at Evolutionary Models 12

2. **Phenomenology and Mechanistics of Dominance** 17
Reiner Veitia and Bruno Bost
Nonlinearity and Dominance 18
Penetrance and Expressivity: The Quantitative Dimension of Dominance and Epistasis 22
The Molecular Biology of Dominance 22
Dosage Effects and the Dosage Balance Notion 30
Dominance and Robustness against Deleterious Effects from a Global Perspective 31

3. **Dominance, Nonlinear Developmental Mapping and Developmental Stability** 37
Christian Peter Klingenberg
Developmental Stability and Its Measurement 38
Developmental Mapping and Nonadditive Genetic Effects 38
Nonadditive Genetic Effects and Developmental Stability 42
Evolution of Developmental Instability 44
Origins of Developmental Noise 45
Perspective: The Challenges Ahead 47

4. **Phenotype and Stochastic Gene Expression: Can the Noise Cause Haploinsufficiency?** 52
Anthony N. Gerber, D.L. Cook and S.J. Tapscott
Deterministic verses Stochastic Gene Expression 52
Haploinsufficiency in a Model of Stochastic Gene Expression 55
Evolutionary Considerations 57

5. **Stochastic Gene Expression: Dominance, Thresholds and Boundaries** 61
H. Frederik Nijhout
A General Model of Stochastic Gene Expression 62
Stabilizing Gene Expression 66
Thresholds and Boundaries 69
Spatially Homogeneous Stochasticity 71
Dominance 74

6. **Mathematical Models of Haploinsufficiency** 76
Indrani Bose and Rajesh Karmakar
Deterministic Model 77
Stochastic Approach 82

7. **Biological Consequences of Dosage Dependent Gene Regulation in Multicellular Eukaryotes** 89
James A. Birchler and Donald L. Auger
Gene Expression in Aneuploids 89
Dosage Compensation 91
Dosage Compensation of the X Chromosome in Drosophila 92
Regulatory Genes Are Responsible for the Inverse and Direct Dosage Effects 93
Haplo-Insufficiency 93
What Is Dominance? 95
How Could Multiple Regulators All Produce a Dosage Effect on the Same Target? 95
Implications for Aneuploid Syndromes 96
Implications for the Genetic Basis of Quantitative Traits 96
The Evolution of Regulatory Hierarchies in Multicellular Eukaryotes 98
Selfish and Cooperative Genes 99
Genomic Imprinting As a Nonmutational Means to Modulate Dosage Dependent Genes 99
How Many Genes Are Important in Evolution? 99
No Need for "Hopeful Monsters" 100
Hybrid Effects and Incompatibilities 100

8. **Clusters of Functionally Related Genes in Eukaryotes, Dosage Balance and Evolvability** 106
Reiner A. Veitia
Some Theoretical Considerations 108
Is Clustering of Interacting Genes Conceivable? 109
Gene Clustering and Dosage Compensation 113

9. Lessons from a Genetic Network about the Causes of Dominance 118
Andreas Wagner
Background of Stoichiometric Network Analysis 119
Main Tasks of Stoichiometric Network Analysis 122
Applications to Robustness 124

10. From Beanbag Genetics to Feedback Genetics: Bridging the Gap between Regulatory Biology and Quantitative Genetics Theory 131
Stig W. Omholt
Characteristics of Current Quantitative Genetic Theory 131
The Implicit Connection between Genetic Dominance and the Feedback Concept 136
Describing and Analysing the Dynamics of an Autoregulatory Locus 138
Relationships between Genetic Dominance and Epistasis 141
Where Do We Go from Here? Outline of a Possible Research Program 144

Index 155

EDITOR

Reiner A. Veitia
Université Denis Diderot/Paris VII
Paris, France
Chapters 2, 8

CONTRIBUTORS

Donald L. Auger
Division of Biological Sciences
University of Missouri
Columbia, Missouri, U.S.A.
Chapter 7

James A. Birchler
Division of Biological Sciences
University of Missouri
Columbia, Missouri, U.S.A.
Chapter 7

Indrani Bose
Department of Physics
Bose Institute
Kolkata, India
Chapter 6

Bruno Bost
Université Paris-Sud
Orsay, France
Chapter 2

Daniel L. Cook
RainTown Biotech
Seattle, Washington, U.S.A.
Chapter 4

Athel Cornish-Bowden
Centre National de la Recherche Scientifique-Bioenergetique et Ingenierie des Proteines
Marseille Cedex, France
Chapter 1

Anthony N. Gerber
Pulmonary Division
University of California, San Francisco
San Francisco, California, U.S.A.
Chapter 4

Rajesh Karmakar
Department of Physics
Bose Institute
Kolkata, India
Chapter 6

Christian Peter Klingenberg
School of Biological Sciences
University of Manchester
Manchester, U.K.
Chapter 3

Vidyanand Nanjundiah
Centre for Ecological Sciences
Indian Institute of Science and Jawaharlal Nehru Centre for Advanced Scientific Research
Bangalore, India
Chapter 1

H. Frederik Nijhout
Department of Biology
Duke University
Durham, North Carolina, U.S.A.
Chapter 5

Stig W. Omholt
Centre for Integrative Genetics
Department of Animal Science
Agricultural University of Norway
Aas, Norway
Chapter 10

S.J. Tapscott
Fred Hutchinson Cancer Research Center
Seattle, Washington, U.S.A.
Chapter 4

Andreas Wagner
Department of Biology
University of New Mexico and
The Santa Fe Institute
Albuquerque, New Mexico, U.S.A.
Chapter 9

PREFACE

The word *dominance*, in the context of genetics, has been used for long time applied to characters or to alleles. A dominant character masks the expression of an alternative form. This loose definition would even apply when these alternatives are not determined by alleles of the same locus. In turn, a dominant allele refers to an alternative version at the same locus. This dual usage has led, as expected, to some confusion and shows how statistics can complement verbal definitions. Mendel, the pioneer of genetics, did not know the bases of the phenomenon of dominance. Nor was he completely certain to look at characters defined by alleles. But the ubiquity of the phenomenon caused him to elevate his observations to the category of laws, that went, unfortunately, unnoticed until they were rediscovered decades later. Today, dominance and recessivity are concepts commonly used and not only by geneticists. Yet a question remains: do we really understand which are the mechanisms of dominance?

Dominance can be considered as a property of either characters or alleles in the living organisms. This must have an explanation in terms of their intrinsic architecture, which has been shaped by natural selection. To what extent dominance is a product of physiology or of evolution has been subject to much debate that is still ongoing. Classical genetics has considered dominance mainly as a result of intra-locus interactions. Indeed, simple mathematical models of positive and negative auto-regulatory circuits provide support to this notion. However, other models show that dominance can also result from the interaction of several loci (epistasis). This points to an intimate connection with epistasis, to such an extent that one can talk about *epistatic dominance effects*. Stemming from this, the word *dominance* in some parts of the text is not used with its classical meaning.

Dominance, from a phenomenological point of view, seems to have a general source: the existence of non-linear relationships between the genotypic and phenotypic values. However, the mechanisms generating these non-linearities are diverse. Models of gene expression, deterministic or stochastic, show the existence of strong non-linearities in the relationship between transactivator concentration and the product of transcription. The beautiful patterns of pigmentation of the butterfly wings are considered here as generated by stochastic processes producing sharp (sigmoidal) boundaries. However, similar events are probably underlying the action of transcription factors involved in developmental processes whose mutations are responsible for many human diseases.

The terms *dominance* and *recessivity* used throughout the text commonly refer to the visible phenotype. Notwithstanding, the consequences of natural selection depend on the dominance at the level of fitness. A mutation can be dominant in respect of the visible phenotype but at the same time be recessive to the wild type in terms of fitness. More important, the direction of allele frequency changes in a population depends on fitness differences of alleles. Although non-linearity can still be the "phenomenological" source of dominance with regard to fitness, the underlying mechanisms are beyond the scope of this book.

Dominance of the wild type phenotype is a special case of robustness against the deleterious effects of mutations. In the extreme, a homozygous gene knockout mutation may have no phenotypic effect. Gene redundancy can be a source of robustness in many but not all cases. Complete obliteration of a gene in a linear pathway considered in isolation is expected to block it completely. In the context of a network, blocking individual steps may have little phenotypic impact, because the information flow can be rerouted. Although this and other types of buffering effects do not concern dominance *sensu stricto*, we have analyzed them for the sake of generality. Moreover, robustness itself, or the epistatic dominance effects mentioned above, are properties emerging from the complexity of the biological systems. Here the cartesian reductionism fails: the whole is not the simple sum of the parts.

The present book is an example of an almost self-organized system. Its main driving force has been the enthusiasm of scholars from all over the world that have agreed to write down their thoughts to enlighten our comprehension of dominance. I hope that this collection of essays will help in the understanding of the bases of mendelian dominance as a pre-requisite to better understand the more complex non-mendelian inheritance. The book relies upon self-contained chapters. They can be considered in the context of a whole, as separate documents, or even as a custom-made printed book. In the end this will save some trees and is in agreement with my perspective that electronic publishing is the future.

Reiner A. Veitia
Université Denis Diderot, Paris

Chapter 1

The Basis of Dominance

Athel Cornish-Bowden and Vidyanand Nanjundiah

Abstract

Dominance is the masking at the level of the phenotype of the presence of one of the two alleles at a diploid locus. We discuss whether it could be an automatic consequence of cellular physiology, as proposed initially by Wright and elaborated by Kacser and Burns, or whether it might be a consequence of natural selection having favoured it in the past, as argued by Fisher. With some exceptions, such as "supply-driven" processes, the physiological explanation is generally valid. At the same time, the extensive pleiotropy of gene action implies that the functioning of a locus can be influenced by many other loci. This means that selection can act on a secondary locus and affect the relationship between alleles at the locus of primary interest. Therefore in principle dominance could have evolved in the manner suggested by Fisher; but when considered in detail, the difficulties with his model imply that the physiological explanation is to be preferred at present.

Introduction

The problem of dominance and recessivity is a classic example of the contrasting ways in which one can view biological phenomena. On the one hand, dominance can be seen as a "here and now" property of a living system. As such, it would seem to require an explanation in terms of the physics and chemistry of the system—and no one doubts that such an explanation must exist. On the other hand, all living creatures have been moulded by evolution; and to the extent that the phenomenon of interest represents an adaptation, this means evolution by natural selection. Then the explanation called for is of an entirely different type: it is to make plausible the assertion that, in the past, the property in question conferred a fitness advantage on individuals possessing it, relative to other individuals who either did not possess it or possessed it in a different degree. It is important to realize that the exercise may result in two seemingly different explanations for the same phenomenon. But, and such is the nature of biology, the fact that one of the two may be right says nothing about whether the other is right or wrong. Part (though not all) of the acrimony in the debates over dominance has resulted from a misunderstanding on this very score, namely, that there are two equally valid levels of explanation in biology.

Mendel[1] observed the distinction between dominant and recessive characters in his classic study of inheritance in peas, but for more than 50 years, even after the importance of this work became recognized at the beginning of the 20th century, this was just an observation—a useful way of classifying characters, perhaps, but not something in need of a mechanistic explanation. The first serious attempt to provide such an explanation was made by Fisher,[2,3] and despite the many attacks on it, his explanation remained generally accepted as plausible, even correct, in the genetics literature until fairly recently (see refs. 4,5). In retrospect, the widespread acceptance of Fisher's ideas appears remarkable in the light of the strong opposition to them from Haldane[6] and Wright,[7,8] both of them authorities as well recognized as Fisher himself. This is especially surprising when seen from a modern perspective[9] in which it seems that Wright's way of looking at dominance is not only easier to understand than Fisher's but also, in the case of pathways of enzyme-catalysed reactions, much closer to being correct.

Mendel used the word dominate to describe an observation that he made on crossing two pure varieties that were distinguishable from each other. In some hybrids the defining trait of one parent was masked by, or recessive to, the corresponding trait of the other parent. The hybrid resembled just one parent, which in consequence was said to dominate the other. Presumably the masking was caused by some interaction between the two strains, not necessarily restricted to the corresponding alleles (as we should be careful to say today). Mendel did not know what caused it, but its ubiquity made him think that he had uncovered a universal functional relationship exhibited by hybrids between pure lines and he elevated the observation to a law. Later workers implicitly assumed that what was happening reflected an intrinsic property of certain alleles relative to others, and this property was referred to as dominance. The difference between dominating and dominance reflects important historical and philosophical differences; see reference 10 for a discussion.

Questions

There is no dearth of evidence testifying to the recessivity of mutant alleles and the dominance of the wild type. That being acknowledged, a few qualifications are in order. Firstly, dominant mutations are not so uncommon either. In particular, many dominant mutations are reported in the data for human genes, amounting to about 25% of all genes that encode proteins with enzymatic activity.[11] Therefore the "law of dominance" has enough exceptions to warrant that they be explained as well. Secondly, dominance is not an all-or-none phenomenon. In general, the fitness of a heterozygous individual is higher than that of the homozygous recessive. At the same time, it is lower than that of the homozygous wild type. Curiously, the fitness effect in the mutant homozygote is negatively correlated with the degree of dominance of the wild-type allele with respect to the mutant allele in the heterozygote. In other words, the more deleterious a mutant in the homozygous state, the closer the fitness of the heterozygote to that of the wild type. Our information on this point is restricted to *Drosophila*[12] and it remains to be seen what the facts are in other species.

The starting point for contemporary discussions of dominance is the assertion that in all organisms the overwhelming majority of mutations are recessive to the prevailing wild type. Though not exactly the same as Mendel's finding, our knowledge of genetics tells us that what he recorded amounts to the same thing (assuming that all his dominant alleles were the ones found in the wild-type phenotype, a point that is unclear). As stated, the assertion has two hidden qualifiers. First, our domain of interest is restricted to diploid organisms and second—a subtle point—the problem to be addressed is why mutant alleles are recessive, not why the wild type is dominant. We list now the questions raised by this assertion in the light of our current understanding of the principles of heredity, the principles of biochemistry, what genes are and how they function. Some of them are interrelated.

1. Is it true? Are there related facts that we need to take into account, or account for?
2. What is it that calls for an explanation, dominance or recessivity?
3. What is the "here and now", in this case genetical or biochemical, explanation?
4. What is the evolutionary explanation? Is it necessary to invoke natural selection?
5. As a corollary of the above, is it likely that natural selection has favoured certain alleles because they were dominant to others? Or are dominance and recessivity likely to have been the outcome of selection for something quite different?

6. Finally, can we arrive at a universally satisfying explanation? (If what we call the wild-type allele pertains to one whose frequency is essentially 1, the necessity for an evolutionary explanation is called into question. Deleterious mutant alleles would be present only as rare, perhaps transient, polymorphisms. A specifically selectionist explanation will have to account for its dominance, not their recessivity when considered individually. The issue has to do with how long an allele remains in a population for selection to act on it, which in turn depends on whether, averaged over all genetic backgrounds, the allele's effect is neutral or not.)

Now we address these questions one by one, adding comments on related issues where necessary.

Framework within Which to Seek Possible Answers

What answers we give to the questions posed earlier depends on our picture of the evolutionary process. If we start by taking it for granted that the overwhelming majority of the population has a wild-type allele at the locus of interest, we need to understand why most mutations are recessive to it. On the other hand, we can argue, as Fisher did, that what we call the wild-type allele itself must have arisen in the past as a novel mutation. In that case, the chances are that it too would have been initially recessive (to the then-prevailing wild type). Then the question is, what has led to its being dominant today? Framing the issue in these terms automatically implies that the problem to be addressed is that of the evolution of dominance per se, even though the failure of a character to be transmitted across generations is the more noticeable property, and hence (one might think) more in need of explanation.

There are two other possibilities. The first is that the mutation that arrived on the scene in the past—obviously an advantageous mutation—must have been dominant to other alleles right from the start (for whatever reason); and we need to understand why. The other is that that mutation may have been recessive initially but has become dominant today as the result of evolution acting, not on dominance as such, but on some other set of traits. It should be stressed that the terms dominance and recessivity commonly refer to the phenotype, hardly ever to fitness. For good reasons, it is taken for granted that most mutations are deleterious. If so, the assertion regarding the dominance of the wild type implies that recessive heterozygotes have the same fitness as the wild type, a conclusion that remains to be tested systematically, as already mentioned in the context of the finding that *Drosophila* shows a significant decrease of fitness in heterozygotes relative to the wild-type homozygote.[12] Unlikely as it seems, one cannot rule out the possibility that most mutations are dominant to the wild type in terms of fitness but at the same time are recessive in respect of the visible phenotype. If this is true in general, the fact that the presence of a mutant allele is masked in respect of some visible trait is neither here nor there as far as evolution is concerned. The masking, meaning dominance of the wild type, would call for a physiological explanation, not an evolutionary explanation in which dominance per se is favoured.

Dominance As a Product of Evolution

Fisher's Model

All discussions under this head must begin with Fisher's theory of the evolution of dominance via modifier genes. He considered that "it is rather a peculiarity of the wild type to be generally dominant than a peculiarity of the mutant to be recessive to the type from which it arose". He believed that in the absence of any mechanism to adjust the results, the intensity of any character would be proportional to the gene dose, so that for genotypes *AA*, *Aa* and *aa* the phenotypes would show 100%, 50% and 0% respectively of normal level of the character. He further considered that this result would be harmful not only to the homozygote *aa* but also to the heterozygote *Aa*. Over many generations, therefore, natural selection would have provided "modifier genes" to bring the heterozygote phenotype to equivalence with that of the normal homozygote. The modifier would act on the heterozygote but would not affect fitness by itself. Selection would act in favour of the modifier at a rate proportional to the product of two factors: the chance that the modifier was present in the heterozygous individual and the fitness advantage caused by its presence.

Difficulties with the Fisher Model

Strictly speaking, if one wishes to explain the evolution of present-day wild type alleles starting from a condition in which they arose as recessives, modifier action should be invoked only in the (original) recessive homozygote—that is, if one wishes to say that the mutation was fully recessive at the beginning. But going by reasonable population sizes and what we know of mutation rates (about 1 in 10^6 to 1 in 10^5 per locus), the population is hardly likely to contain individuals that are homozygous for the mutant allele right from the start. Even if we say that the modifier acts on the heterozygote, which amounts to assuming that the mutation had a weak dominant effect at the beginning, the pool of individuals on which selection can act remains miniscule because the mutation rate is extremely low. The number of relevant individuals in reasonably sized interbreeding groups could be so small that evolution for dominance, besides being slow, would run the risk of being overwhelmed by selection acting on the modifier gene itself.[7,8] Fisher discounted this possibility: he thought that populations would always be large enough and that one could envisage modifiers that had no direct effect on their own. It can be shown that if the modifier evolves because of a direct effect on fitness, selection for dominance need not be hampered by the fact that the mutation rate, or the pool of heterozygotes, is small.[13]

Haldane[6] also criticized Fisher's theory on grounds somewhat similar to those of Wright. He put forward an alternative explanation in which dominance had coevolved as a by-product of selection in favour of buffering the genotype against environmental or genetic perturbations. Selection for buffering would gradually change the phenotypic value of the mutant heterozygote to the same level as in the wild-type homozygote. Much later, Haldane[14] drew on his investigations into the evolution of industrial melanism (in which the melanic form is both advantageous and dominant) and suggested that dominance might have evolved from a low level, conceivably by a Fisher-type process, concomitantly with the spread of the allele. Note that in Haldane's scheme the pool of heterozygotes is large and keeps increasing. However, in common with Fisher's model both his proposals share the assumption that selection acts on the heterozygote.

The criticisms made by Wright and Haldane were on the grounds of plausibility. Besides that, there are two problems with Fisher's model, both based on observational grounds. The first is that it does not explain why, in the heterozygous state, the presence of a strongly deleterious allele is not as apparent as the presence of a weak allele. In other words, the model cannot account for the observed negative correlation between the degree of dominance and the fitness of the mutant homozygote.[15] The second problem is more serious. Fisher included in his explanation of dominance a feature that almost precluded any possibility of experimental verification, namely that the selection of modifier genes would require an enormous number of generations before the heterozygous phenotype became indistinguishable from that of the normal homozygote. This would mean that one could not study successive generations for a sufficiently long time to observe the gradual evolution of dominance.

Nonetheless, the model does lead to a specific prediction that is open to testing in an appropriate organism. It is obvious that concepts such as dominance and recessivity have no meaning in haploid organisms; in addition, less obviously, Fisher's model implies that modifier genes for dominance, and hence the evolution of dominance, cannot occur in haploid organisms, and no dominance should be observed in a haploid organism that passes through occasional diploid or polyploid generations. *Chlamydomonas reinhardtii* is such an organism: it exists normally as a haploid organism, with one nucleus per cell, and with no more than a single copy of each chromosome, but it also has diploid or polyploid generations only occasionally. These occur too infrequently for the organism to experience the consequences of heterozygosis to drive the selection of suitable modifier genes. If Fisher's hypothesis is correct, therefore, one should see no sign of dominance in diploid generations of *C. reinhardtii.*

What is observed is quite different. Orr[16] examined numerous mutations in diploid cells and found that the great majority were recessive, exactly as would be expected from the theory of Kacser and Burns[9] developed from Wright's ideas, but inconsistent with that of Fisher. Obviously, if a property that can only evolve if it is selected in diploid (or polyploid) cells proves to be exactly the same in a species that is nearly always haploid, an explanation for that property that requires the long-term persistence of diploidy cannot be a universal explanation. Orr noted that his observations disposed not only of Fisher's model but also of the two proposed by Haldane.[6] Haldane's models also required selection in heterozygotes, and therefore could not explain why mutant genes are as likely to be recessive in the very rare heterozygotes of a principally haploid species like *C. reinhardtii* as they are in species that are always diploid.

Wright's Theory: Dominance As a Physiological Phenomenon

We have seen how Wright[7] considered that the selection pressure would be too weak for Fisher's explanation to work, and later[8] he pointed out that the modifier genes were themselves subject to mutation, and thus in need of their own modifier genes to protect them from the effects of such mutations, and so on, with obvious possibilities of infinite regress. He argued that an explanation for dominance should be sought in mechanistic terms, meaning in terms of immediate physiology. Even with care to avoid reading more into Wright's words than he intended, it is clear that his statement that "the curve expressing the relation of *the product* to *enzyme amount* is a hyperbola, asymptotic at its upper limit. Doubling the quantity of *enzyme* will less than double the amount of *product*" (in which the italicized words and phrases replace algebraic symbols in the original) contained two important ideas. First, there is nothing here that can be interpreted as a reference to genes, so we are dealing with an explanation at a physiological or biochemical level; second, he recognizes at the outset that biochemical responses to enzyme levels are nonlinear.

As we shall see, this makes his explanation of recessivity far closer to current ideas than anything that Fisher or Haldane wrote, even though Haldane was much better informed about biochemistry in general and enzymes in particular than either of his rivals: his book,[17] written at about the time of the controversy, is a classic account of enzymes that is still worth reading for its essential insights more than 70 years afterwards.

Nonetheless, it was Fisher's ideas that were generally accepted, not those of Haldane or Wright. As late as 1966, Wright's view of recessivity as a physiological phenomenon was rejected as impossible;[18] and in 1981 an important book[19] by an author justifiably famous for the clarity and cogency of his writing could still contain some obscure pages presenting a vague exposition of Fisher's ideas barely more intelligible or convincing than Fisher's own. In the same year, however, Kacser and Burns[9] provided a fully modern version of the physiological theory, in which they showed that dominance and recessivity follow automatically from the known kinetic behaviour of enzymes in isolation and when embedded in metabolic pathways.

Effect of Enzyme Activity on Metabolic Flux

For almost any enzyme in isolation, for example in a spectrophotometric assay, the rate of reaction at specified concentrations of substrates, products and any inhibitors or activators present in the mixture is proportional (at least approximately) to the concentration of the enzyme: if the activity of enzyme is doubled, the rate is doubled. Exceptions occur if the enzyme exists in several different states of association with different degrees of catalytic activity,

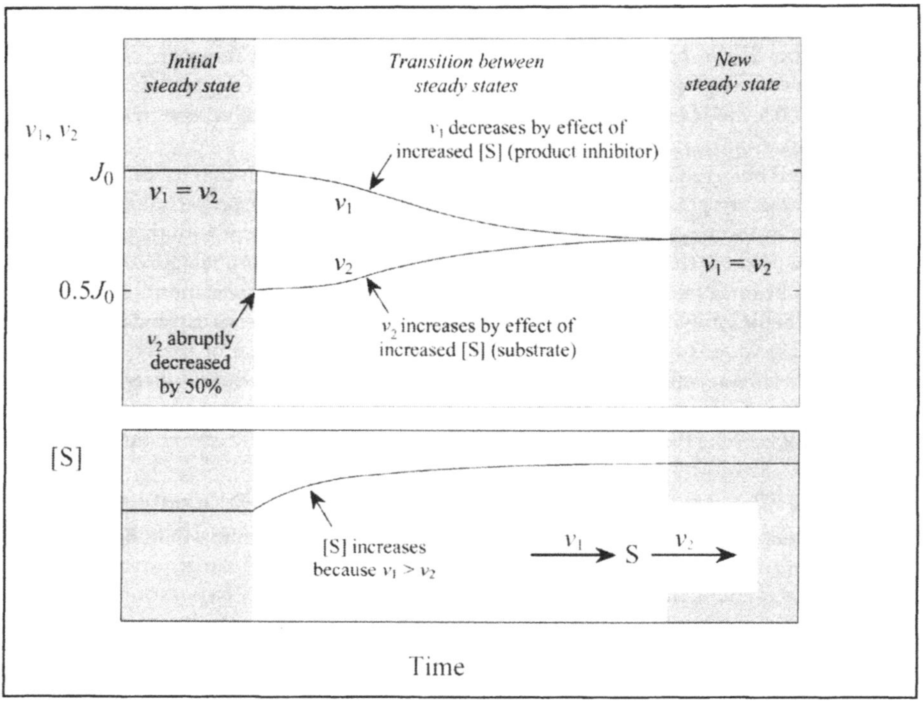

Figure 1. Effect of an abrupt change in enzyme activity in a system of two enzymes in steady state. The system considered is shown in the inset at bottom-right. In the initial steady state the two rates v_1 and v_2 are equal to one another at some arbitrary value J_0. If the activity of the second enzyme is abruptly decreased by 50%, this must produce a corresponding abrupt decrease in v_2 to $0.5J_0$, but the system is then no longer in steady state because $v_1 > v_2$, and so the intermediate S is being released faster than it is being consumed. Its concentration [S] must therefore increase, with two effects: as S is the product of the first enzyme its increased concentration increases the product inhibition and so v_1 decreases; at the same time it is the substrate of the second enzyme, and so by the usual effects of substrates on enzymes the rate v_2 increases. Eventually a new steady state is reached in which the two rates are again equal, at a value smaller than J_0 but larger than $0.5J_0$.

and the experiment is done in the concentration range where the degree of association varies. Other exceptions arise if the enzyme is unstable at high dilution, a problem common enough in kinetic studies to oblige experimenters to take precautions to handle it properly. However, it is an artificial problem in the physiological context, because the high enzyme dilutions commonly used to make steady-state rates slow enough to study conveniently in the spectrophotometer are themselves unphysiological. Nonetheless, the usual proportional dependence of rate on enzyme concentration observed in steady-state kinetic experiments is misleading as a guide to behaviour in the cell, because in the cell the concentrations of substrates, products and so on are not constants fixed by an external agent, the experimenter, but variables that depend on the activities not only of the enzyme of interest but of all the enzymes in the system. The product of almost any enzyme is also the substrate of one or more other enzymes; the substrate of almost any enzyme is also the product of one or more other enzymes.

Consider now what will happen in response to a two-fold decrease in the activity of the second of two enzymes in a pathway that constitutes the whole of a two-enzyme system (Fig. 1). The immediate effect will be the loss of any steady state that existed before the change, because the first enzyme will initially continue to work at the original rate, with the second working at only half that rate. The common intermediate, the product of the first enzyme and substrate of the second, is thus no longer in steady state, because it is being produced faster than it is being consumed: its concentration must therefore increase, but that has its own effects, increased inhibition of the first enzyme and increased saturation of the second. Eventually these will produce a new steady state in which the rates are again balanced, but notice that the increased inhibition of the first enzyme and increased saturation of the second mean that the new steady-state rate must be less than the original rate but greater than half the original rate. In other words the effect of decreasing the activity of the second enzyme is less than proportional. This result is general, applying to changes more or less than two-fold, to changes in the activity of the first enzyme as well as to that of the second, and to systems with any number of enzymes greater than one. In all these cases the result of changing any enzyme activity is a less-than-proportional change in the steady-state flux through the system.

The original analysis was made by Kacser and Burns[20] before they analysed dominance and recessivity, and is now discussed in textbooks,[21,22] so here it is sufficient to give the main result. The effect on the metabolic flux of any enzyme activity can be expressed as a flux control coefficient, defined as the logarithmic derivative of the flux J with respect to a perturbation p divided by the logarithmic derivative of the enzyme activity v_i with respect to the same perturbation p:

$$C_i^J = \frac{\partial \ln J}{\partial \ln p} \Big/ \frac{\partial \ln v_i}{\partial \ln p} = \frac{\partial \ln J}{\partial \ln v_i} \qquad (1)$$

The anonymous perturbation p is introduced to avoid the mathematical looseness in the simpler form at the right derived from the fact that v_i is not a true independent variable of the system; the simpler form is acceptable as long as it is remembered that an external perturbation is implied even if not specified. Moreover, to the extent that enzyme activities are proportional to enzyme concentrations one can regard the enzyme concentration as the perturbation p: this was done in the original paper of Kacser and Burns,[20] but in more recent work most authors have preferred to avoid an assumption that is not only unnecessary but also, more seriously, was responsible for a widespread misconception that metabolic control analysis deals only with changes in activity brought about by changes in enzyme concentration. By a more rigorous analysis along the lines of what was done above for a two-enzyme system, one can readily prove the fundamental property of flux control coefficients expressed by the following equation:

$$\sum_{i=1}^{n} C_i^J = 1 \qquad (2)$$

in which n is the number of enzymes in the system. This is called the summation property for flux control coefficients, and expresses the fundamental idea that flux control is shared among all the enzymes in the system. As even the simplest real cell contains hundreds of different enzymes the average share must be very small. More realistically, perhaps, noting that most enzymes have a very small share of the control of flux through pathways other than those in which they occur, one can say that most of the control of the flux through any pathway is shared among the enzymes of that pathway. This still implies an average share of no more than 20% for most pathways, and often less.

There is a complication in this argument that needs to be made explicit. In referring to shares we are tacitly assuming that all the control coefficients in the sum shown in eqn. (1) are positive, but that is not strictly true. The flux control coefficient of an enzyme for a flux through a part of the metabolic network not in series with the reaction that it catalyses can certainly be negative, and often is. However, in nearly all real cases such negative flux control coefficients are small enough to have little effect on the general argument. Instead of saying that control of flux through any pathway is exactly shared among all the enzymes of the system we must say that most of it is shared among the enzymes of the pathway. This statement is not very different from what we said in the previous paragraph.

Before leaving the question of negative control coefficients, we should note that these can be defined for other system variables as well as fluxes; for example, the effect of enzyme activities on any metabolite concentration [S] can be expressed by concentration control coefficients with definitions very similar to that in eqn. (1). The resulting quantities are often not merely negative but large in magnitude as well, and the summation property corresponding to eqn. (2) has a right-hand side of zero, not unity. In such a case the idea of sharing loses all meaning, so we should emphasize that the idea of sharing flux control comes not merely from the existence of a summation property but from two additional points, one theoretical and general, the other from observation: first the sum is unity; second, the individual elements in the sum are mostly positive and, when negative, are normally small. On a preliminary view the summation property in eqn. (2) already provides the explanation of dominance and recessivity, as it suggests that changing any enzyme activity will produce a much less than proportional change in metabolic flux, and hence a much less than proportional change in metabolic output, such as the amount of a pigment produced by a pea plant. However, that is too simple, as it applies only to small—strictly infinitesimal—changes in activity, whereas we need to consider the effects of changes in activity of the order of two-fold. For increases in activity the naive view remains valid, but for decreases, which are more relevant to heterozygotes in diploid organisms, it does not, because any flux control coefficient typically increases when the enzyme activity is decreased, eventually reaching a value of 1 when the activity is close to zero.

Both simple models and experimental observations in numerous cases indicate that for finite changes in activity the curve relating metabolic flux to the activity of any enzyme in the pathway resembles a rectangular hyperbola. In the simplest case where all enzymes operate far enough below saturation for the kinetics with respect to their substrates to be of first-order the curve is exactly a rectangular hyperbola; it remains quite similar to one in the more realistic case where some or all of the enzymes operate in a range with detectable saturation. In Michaelis–Menten terms this implies substrate concentrations of the order of the relevant Michaelis constants. What eqn. (2) now means is that most enzymes are located near or on the flat part of the hyperbola in the state corresponding to the normal homozygote. If any particular enzyme happens to have a flux control coefficient close to unity—unusual, but not impossible — then all the others in the pathway need to move further to the right along the curve.

Consider now the expected metabolic fluxes J_{AA}, J_{Aa} and J_{aa} for the normal homozygote, heterozygote and abnormal homozygote respectively of the gene for an enzyme normally located in the typical position on the hyperbola, as shown in Figure 2. It is immediately evident not only that $J_{AA} \approx J_{Aa} >> J_{aa} = 0$, but also, more important, that this relation will survive even quite large variations in the assumption about where the normal enzyme lies on the curve: it can be displaced any distance to the right without invalidating the relationship, and significantly to the left also.

This, in essence, is the explanation of dominance and recessivity proposed by Kacser and Burns (1981). It may be objected that it predicts only that the heterozygote phenotype will be more similar to the normal homozygote phenotype than to the abnormal homozygote phenotype, not that it will be identical to it. This is true, but in practice the difference would not be likely to have been noticed in Mendel's observations, or indeed in most observations of gross phenotypes since then. In summary, therefore, the theory of Kacser and Burns (1981), which was an elaboration and extension of that suggested by Wright (1934), is the only one currently available that provides a satisfactory explanation of dominance and recessivity.

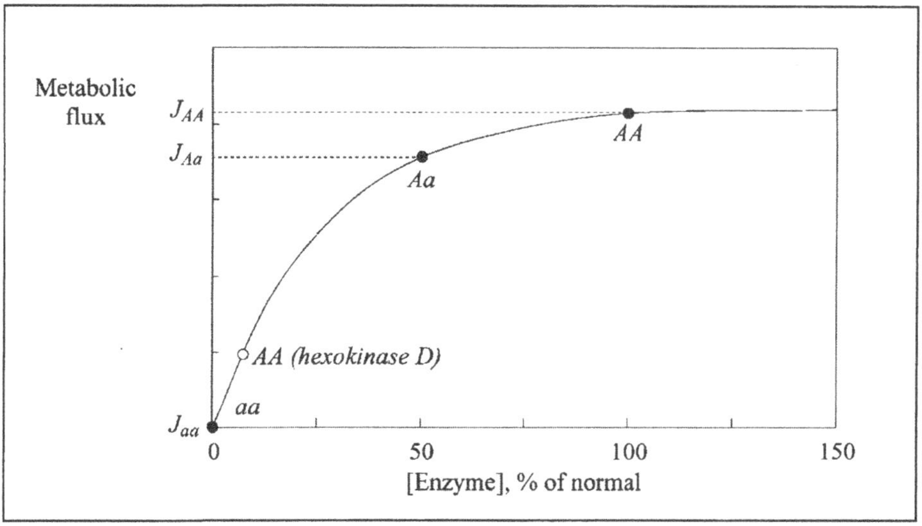

Figure 2. Typical dependence of flux through a metabolic pathway on the activity of any enzyme in the pathway. The curve resembles a rectangular hyperbola, but is not exactly one except under rather special circumstances. Most enzymes have low flux control coefficients, which means that they are typically located on a part of the curve where the slope is small, for example at the point labelled *AA*. If this represents the activity of an enzyme in the normal homozygote, then the point *aa* represents an abnormal homozygote in which the enzyme activity is entirely lacking, and the point labelled *Aa* represent the heterozygote. The corresponding phenotypes are related to the fluxes through the pathway, and can be estimated as the ordinate values labelled J_{AA}, J_{aa} and J_{Aa}. Note that the heterozygote phenotype is much closer to that of the normal homozygote than to that of the abnormal homozygote. The point labelled *AA (hexokinase D)* refers to a special case discussed at the end of the chapter.

Effect of High Saturation of Enzymes

Before the theory of Kacser and Burns[9] is accepted as the full explanation of dominance and recessivity there are some complications that need to be dealt with. The first is the question of whether it is necessarily true that the curve for dependence of metabolic flux on the activity of any enzyme resembles a rectangular hyperbola as closely as the one illustrated in Figure 2. In fact it is not necessarily true, because it is possible to devise models in which the normal steady state is one in which all of the enzymes in a sequence are close to saturation with their substrates,[23] and in such a case the curve relating flux through the pathway to the activity of any enzyme resembles the one shown in Figure 3. This differs from a rectangular hyperbola in having much more extended domains of first- and second-order dependence, with a more abrupt transition between them, and the enzymes in the homozygote are typically perched as shown near the transition region. In such a case the relationship embodied by eqn. (2) remains true; i.e., it remains true that very small changes in the activity of any enzyme produce even smaller effects on the flux. However, the effects of decreases of the order of 50% are quite different from what is seen in Figure 2. Now the heterozygote phenotype is very close to half-way between the homozygote phenotypes, as it was assumed to be in early models. We must emphasize, that this was a pathological model constructed to demonstrate that dominance did not follow necessarily from the summation property. It was not intended as a realistic model of how real systems would normally behave, and Kacser[24] was justified in entitling his response to it "Dominance not inevitable but very likely".

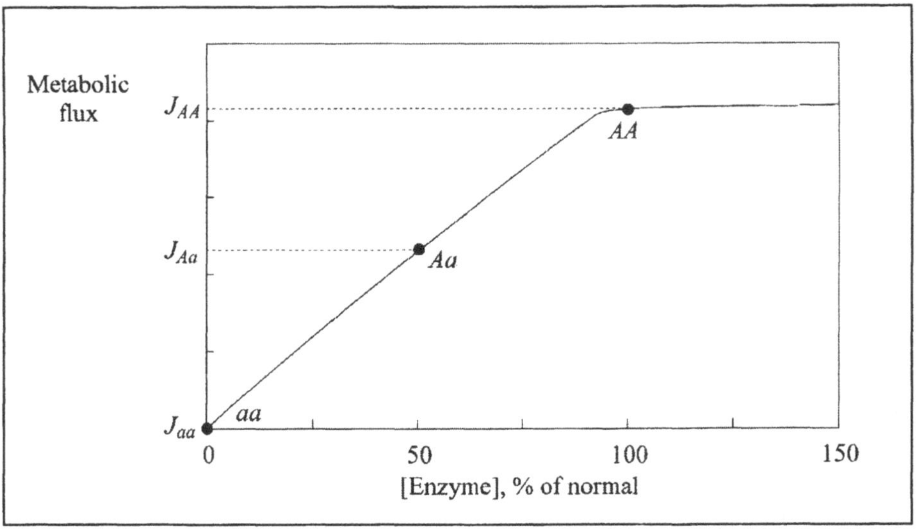

Figure 3. Dependence of flux on the activity of any enzyme when all enzymes are close to saturation with their substrates. Although it remains true that an enzyme in the normal homozygote is typically located at a point where the slope is small, the slope in this case changes sharply if the activity is decreased, even by a small amount, and for this model the phenotypes of the heterozygote is close to the midpoint between those of the homozygotes.

Even in the absence of any selection, assigning kinetic constants entirely at random to the enzymes in a metabolic pathway would be very unlikely to produce a set generating behaviour similar to that in Figure 2. If a primitive organism happened to possess such a pathway one could expect its kinetic characteristics to be varied by natural selection towards a more typical behaviour. That is because the model proposed is not only pathological in the modelling sense; it would also be pathological in a living organism, because any enzyme working close to saturation represents a danger for the organism.[25] For an enzyme operating far below saturation, for example at $0.1V$, 10% of the limit, with a substrate concentration equal to $0.1K_m$, the rate is correspondingly far below the limit, about $0.09V$ for the numerical case considered. In consequence a sudden increase in the flux to the pathway brought about by increased activity of the upstream enzymes would present little problem to the enzyme: doubling the flux to $0.18V$ would imply an easily sustainable increase in substrate concentration to $0.22K_m$, barely more than doubled.

Matters are quite different for an enzyme operating at $0.95V$: a mere 5% increase in flux is now sufficient to bring it to a region where no steady state exists, a potential catastrophe. It is reasonable to suppose, therefore, that if any pathway existed in which all enzymes were flux-controlling for non-infinitesimal decreases in activity then natural selection would act to moderate such behaviour. Thus we cannot entirely eliminate natural selection from the discussion of dominance and recessivity, but it is a much more physiological and intelligible form of natural selection than what was envisaged by Fisher. Grossniklaus et al[26] examined a broader range of models than that considered by Cornish-Bowden,[23] and found that even when enzymes are not close to saturation substantial deviations from Michaelis–Menten kinetics, especially in cases with a large degree of positive cooperativity, can cause some enzymes to have large flux control coefficients, so that mutations in these enzymes would not be recessive.

Dominant Genes in Humans: An Exceptional Species?

As we have seen, the controversy over the explanation of dominance arose from the observation that mutant genes were normally recessive. Until comparatively recently it appeared that this was quite general, observed in all diploid and polyploid species. Remarkably, however, the human species appears to be exceptional, with a high proportion of mutant genes reported to be dominant. How is this to be explained? Must we return to a world view in which the human species is unique unto itself, and quite separate from animals, plants, fungi and bacteria? Clearly not, but the observation appears genuine, and needs a plausible explanation. The most likely one is that this difference between humans and animals (like most differences between humans and animals) owes more to the way in which they are studied than to any inherent differences in the underlying properties. The idea of a symptomless disease would appear absurd in relation to animals or plants, but is applied in all seriousness to human conditions like maturity-onset diabetes of the young, or "MODY" (infection with human immunodeficiency virus may be a better example, but we shall not enter into that controversy here). In such cases there is no sickness, but biochemical tests indicate abnormal levels of certain metabolites that indicate that a disease is likely to develop later. Applying this idea to other organisms, and bearing in mind the behaviour suggested by Figure 2, it seems clear that Mendel could have distinguished between homozygotes and heterozygotes of pea plants producing green seeds if he had used modern instruments to measure the exact amounts of green pigment in his samples, rather than relying on his eye to distinguish between green and yellow seeds.

In general, therefore, the apparent difference in heterozygote characteristics between humans and other animals is best explained by the greater detail in which humans are subjected to biochemical tests. Nonetheless, there is at least one example of a dominant gene—first recognized in humans, though probably not a special property of humans—that can serve as an archetype to illustrate when such genes are to be expected. This will be considered next.

The Hexokinase D ("Glucokinase") Gene in Humans

Hexokinase D is the isoenzyme of hexokinase characteristic of the vertebrate liver (see ref. 27), and like any hexokinase it catalyses the phosphorylation by ATP of glucose and other hexoses, such as fructose and mannose. However, it differs from the other isoenzymes found in mammals in several respects (though not in specificity, as implied by the misleading name "glucokinase" that is often used for it): it is half-saturated at much higher glucose concentrations, of the order of the glucose concentration in the blood; it does not follow Michaelis–Menten kinetics with respect to glucose but shows positive cooperativity instead; it is unaffected by its product glucose 6-phosphate at physiological concentrations. All of these properties fit it for the role of helping to maintain a constant glucose concentration in the blood: when this increases above the normal level of 5 mM the activity of hexokinase D rises steeply and glucose is stored in the liver as glycogen; when the blood-glucose concentration falls hexokinase D decreases its activity and other enzymes catalyse the mobilization of the glycogen.

These characteristics not only set hexokinase D apart from the other hexokinase isoenzymes; they also set it apart from most enzymes known to have important roles in metabolic regulation, because they imply that conversion of blood glucose to liver glycogen is a supply-driven process, whereas most of the pathways that have been well studied are best understood as demand-driven processes.[28] This does not affect the general truth of the summation property expressed by eqn. (2), which is derived without reference to the properties of the specific enzymes considered, but it does affect the way the control is distributed among the terms in the sum. In a typical demand-driven pathway properties such as feedback inhibition act to transfer control away from the supply part of the pathway towards the reactions that consume the end-product. However, the liver has little demand for glucose 6-phosphate, such little as it has being amply satisfied by the other hexokinase enzymes,[29] and control of glycogen synthesis is concentrated almost entirely in hexokinase D.[30] This concentration of flux control in hexokinase D means that it is located much further to the left on the curve in Figure 2 than a typical enzyme, as indicated by the point labelled *AA (hexokinase D)*, and that the argument used previously to conclude that the heterozygote phenotype ought to resemble that of the normal homozygote does not apply to hexokinase D. Instead, heterozygotes for hexokinase D should be much less efficient than homozygotes for maintaining a correct blood-glucose concentration, and should have a readily recognizable hyperglycaemia. In accordance with this expectation, young heterozygotes display a mild hyperglycaemia that is taken as the form of type II (non-insulin-dependent) diabetes known as maturity-onset diabetes of the young, an example of a "symptomless disease", as already noted. Later, in adulthood, such heterozygotes usually show increased hyperglycaemia and the pathological symptoms of type II diabetes, thus providing a genuine example of a disease with dominant transmission.[31] However, it should be clear from this discussion that the dominance is here a consequence of the special properties of hexokinase D, and does not support the notion that dominant mutations are any more common in humans than in other species. It would, in fact, be very surprising if hexokinase D mutants in animals such as dogs and rats did not prove to behave the same way as in humans if the genetics of type II diabetes were studied in as much detail in these animals as it has been in humans.

Another Look at Evolutionary Models

What we have said so far seems to settle the issue of dominance and recessivity. It would appear that they are straightforward consequences to be expected from the biochemistry of metabolic networks. The networks must be a product of evolution, of course, but dominance would not be something that has come about as a direct result of natural selection acting in its favour. For all these reasons Fisher's original model no longer seems credible as a working hypothesis. However, one should be wary of concluding that evolutionary models for dominance should not be taken seriously.

There are many reasons for saying this. For one thing, even when a DNA sequence or gene product seems to be "responsible for" a particular trait (as inferred by mutations in the gene or by a knowledge of the relevant pathways), the same trait can also be influenced by many other genes. In short, modifier loci are ubiquitous. Sometimes modifiers are referred to as "minor" genes in contrast to the "major" gene (see ref. 32). Ironically, considering his long-standing opposition to the Fisher model, Wright was amongst the strongest advocates of what he called "generalized pleiotropy", the concept that single genes influence many traits. Pleiotropy is accepted as a truism today, and proteome data show that point mutations typically affect expression levels for hundreds of proteins.[33,34] It follows that depending on the locus at which we direct our attention, many—perhaps almost all—other loci can be thought of as modifiers. Extensive field studies led Sheppard and Ford[18] to conclude that modifier loci could influence dominance and that dominance modification can evolve in the wild. Experiments have shown that artificial selection for dominance modification works quite successfully.[35-37]

Pertinent to the question of what has happened during evolution, Doebley et al[38] investigated the genetic differences between modern maize and teosinte, which is believed to resemble the ancestral species. A number of alleles from two of the five major regions of the genome that were so identified were transferred by means of appropriate crosses from maize to teosinte. Their finding is consistent with the following interpretation: the dominance that is seen in maize breaks down when the same genes are transferred to teosinte because it has a different genetic background—lacking, presumably, the right set of modifier loci.

So the question is not so much about whether dominance can evolve in the manner suggested by Fisher as whether it does. As already pointed out, dominance could have evolved as an incidental consequence of the evolution of something else, but that is not the same thing as the evolution of dominance "for its own sake" (as Fisher meant). The contention of Bagheri-Chaichian and Wagner[13] is that the "something else" is simply the result of direct selection acting on the modifier. But there may be conditions in which Fisher's model is not so unrealistic. Otto and Bourguet[39] discuss ways in which mutant heterozygotes could be present at far higher frequencies than they would be based solely on a consideration of mutation-selection balance, thereby resurrecting variants of Haldane's proposal[14] mentioned above. They put forward two schemes, and in both of them heterozygous individuals are present at an appreciable frequency. In one, heterozygotes are maintained by overdominance (the heterozygous state has a higher fitness than both homozygotes). In the other scheme, a patchy environment with inter-patch migration (at rates much higher than typical mutation rates) leads to selection in favour of dominance modification even when the effect of the modifier on heterozygous fitness is negative in one patch. The essence of both their schemes is a spatially varying coefficient of natural selection. As Otto and Bourguet[39] point out, the evolution of pesticide resistance satisfies the requirements of their model: resistance is advantageous in areas with high pesticide levels but disadvantageous elsewhere. The classic example of genetic polymorphism being maintained by heterozygote advantage (i.e., overdominance) is that of sickle cell anaemia in a malarial environment. It would be worth investigating whether dominance modifiers have influenced phenotypes in this case.

Concluding Remarks

Dominance and recessivity bear comparison with other situations in which the phenotype of a diploid organism is insensitive to the functional absence of one of the two copies of a gene or set of genes whose activity is known to influence it. Dosage compensation for sex-linked genes is a well known example: where the male is the sex with heteromorphic sex chromosomes (as in flies and worms), the single X chromosome in males seems to be "as good as" the two X chromosomes in females. (Eutherian mammals achieve the same result by randomly inactivating one of the two X chromosomes in each cell of the female.) Both evolutionary and nonevolutionary arguments have been advanced to explain dosage compensation, and they resemble the arguments for dominance and recessivity that we have been discussing.[40]

Haploinsufficiency is the obverse of what we have been discussing: it refers to a situation in which the loss of one copy of a gene leads to detectable effects and offers a route to understanding dominant mutations.[41,42] One way of looking at both dosage compensation and haploinsufficiency is that they show how important it is to maintain a proper balance between the levels of various gene products and what can go wrong when the balance is disturbed. "Balance" arguments can be of two kinds. They can start by taking it as given that males and females must have, for other reasons, different copy numbers of X-linked genes. Then they can invoke selection for modifiers that redress this imbalance—the implicit assumption being that most X-linked genes encode products that perform roles that are independent of the sexual phenotype (see ref. 43). On the other hand, considerations of optimal functioning of a genetic or biochemical network and need not invoke selection for balance per se, but balance can be a "system" or "network" property.[44,45]

The concept of canalization or buffering[46] is more general than that of balance, and therefore all the more relevant for the phenomena of dominance and recessivity. The reasoning goes as follows: if natural selection demands well adapted phenotypes, genotypes must be capable of tolerating—i.e., they must be buffered with respect to—genetic and environmental stresses; and the absence of one of the two wild-type copies of a gene can be thought of as a form of genetic stress.[47,48] Forsdyke[49] has proposed a model for the evolution of dominance that depends on selection for buffering the effects of heat shock. A point not considered by us is that phenotypes can depend on reactions whose features differ in essential respects from those of simple enzyme-catalysed pathways operating far from saturation. For example, transcriptional cascades are at the heart of genetic regulatory pathways, and the relevant kinetic steps resemble more a series of zero-order reactions than successive Michaelis–Menten type reactions. As we have seen, in such a situation mutations would be more likely to be dominant than recessive.[23] In certain cases the phenotype of interest is the level of a product and not the flux through the pathway that results in the product (for example, a stable pigment which colours a part of the body). Once again the expectation would be that a loss-of-function allele is dominant to the wild type. More generally, the fact that the genotype–phenotype map is complicated[34] makes it necessary to develop generic models for dominance and recessivity, for example models that depend solely on features such as the nonlinearity of the map. Gilchrist and Nijhout[50] have taken the first steps in this direction, and it is gratifying that their analysis can account as well for dominant mutations as for recessive ones.

In the light of Orr's observations on *Chlamydomonas*,[16] the outcome of recent laboratory experiments with the yeast *Saccharomyces cerevisiae* is curious. Zeyl et al[51] were trying to compare the fitnesses of haploid and diploid yeasts that were maintained asexually in shaken cultures via serial transfer for over 2000 generations. Adaptive mutations arose in the course of the experiment. The suite of adaptive mutations that arose in one haploid and one diploid line each were tested for dominance or recessivity (in the diploid condition, of course), with, unusually, reference to fitness as the relevant measure of phenotype. It turned out that the mean dominance coefficient in the haploid genotypes was 0.20 whereas that in the diploid genotypes was 0.75, with some exhibiting overdominance. In other words, among those mutants that were positively selected during the course of the 2000 generations (or were not selected against), those that arose in a haploid background were on the whole recessive and those that arose in a diploid background were on the whole dominant, both features referring to observations made at the end of the experiment. The first observation seems to be in line with Orr's finding and the inferences drawn from it, both of which we have discussed. But the second is, to say the least, unexpected. We need to wait for more data to see whether it is a trivial consequence of selection for dominant mutations under the conditions of the experiment or whether it indicates an essential difference in the spectrum of mutations that occur in haploid and diploid yeasts.

In conclusion, both theory and experiment indicate that the parsimonious explanation for dominance and recessivity has to do with genetics and cellular physiology alone, not on evolution per se; the nature of the explanation may differ from one situation to another. If we adopt this as our stand, the fact that we can carry out selection for dominance modification, and the reason why the usual pattern of dominance can break down when a gene finds itself in a new genomic and cytoplasmic environment, may have to do with the evolution, for quite different reasons, of complex genetic networks. The alternative is to say that nonevolutionary explanations account for most cases of dominance but there are also cases in which evolution has selected for the trait of dominance.

Acknowledgement

We thank María Luz Cárdenas and H. Sharat Chandra for useful comments on this chapter.

References

1. Mendel G. Versuche über Pflanzenhybriden. Verhandlung des naturforschenden Vereines in Brünn 1866; 4:3-47.
2. Fisher RA. The possible modification of the response of the wild type to recurrent mutations. Amer Nat 1928; 62:115-126.
3. Fisher RA. Two further notes on the origin of dominance. Amer Nat 1928; 62:571-574.
4. Nanjundiah V. Why are most mutations recessive? J Genet 1994; 72:85-97.

5. Porteous JW. Dominance—one hundred and fifteen years after Mendel's paper. J Theor Biol 1996; 182:223-232.
6. Haldane JBS. A note on Fisher's theory of the origin of dominance and a correlation between dominance and linkage. Amer Nat 1930; 64:87-90.
7. Wright S. Fisher's theory of dominance. Amer Nat 1929; 63:274-279.
8. Wright S. Physiological and evolutionary theories of dominance. Amer Nat 1934; 63:25-53.
9. Kacser H, Burns JA. The molecular basis of dominance. Genetics 1981; 97:639-666.
10. Falk R. The rise and fall of dominance. Biol Philos 2001; 16:285-323.
11. Jimenez-Sanchez G, Childs B, Valle D. Human disease genes. Nature 2001; 409:853-855.
12. Simmons MJ, Crow JF. Mutations affecting fitness in Drosophila populations. Ann Rev Genet 1977; 11:49-78.
13. Bagheri-Chaichian H, Wagner GP. Evolution of dominance through incidental selection. SFI working paper at http://www.santafe.edu/sfi/publications/wpabstract/200211064. 2002.
14. Haldane JBS. The theory of selection for melanism in Lepidoptera. Proc Roy Soc Lond B 1956; 145:303-306.
15. Charlesworth D. Evidence against Fisher's theory of dominance. Nature 1979; 278:848-849
16. Orr HA. A test of Fisher's theory of dominance. Proc Natl Acad Sci USA 1991; 88:11413-11415.
17. Haldane JBS. Enzymes. London: Longmans Green, 1930.
18. Sheppard PM, Ford EB. Natural selection and the evolution of dominance. Heredity 1966; 21:139-147.
19. Dawkins R. The Extended Phenotype: The Long Reach of the Gene. Oxford: Oxford University Press, 1981.
20. Kacser H, Burns JA. The control of flux. Symp Soc Exp Biol 1973; 27:65-104.
21. Fell D. In: Understanding the Control of Metabolism, London: Portland Press, 1997.
22. Cornish-Bowden A. In: Fundamentals of Enzyme Kinetics. 3rd ed. London: Portland Press, 2004.
23. Cornish-Bowden A. Dominance is not inevitable. J Theor Biol 1987; 125:333-338.
24. Kacser H. Dominance not inevitable but very likely. J Theor Biol 1988; 126:505-506.
25. Atkinson DE. Regulation of enzyme function. Annu Rev Microbiol 1969; 23:47-68.
26. Grossniklaus U, Madhusudhan MS, Nanjundiah V. Nonlinear enzyme kinetics can lead to high metabolic flux control coefficients: Implications for the evolution of dominance. J Theor Biol 1996; 182:299-302.
27. Cárdenas ML. "Glucokinase": Its regulation and role in liver metabolism. Austin: RG Landes 1995.
28. Hofmeyr J-HS, Cornish-Bowden A. Regulating the cellular economy of supply and demand. FEBS Lett 2000; 476:47-51.
29. Reyes A, Cárdenas ML. All hexokinase isoenzymes coexist in rat hepatocytes. Biochem J 1984; 221:303-309.
30. Agius L, Peak M, Newgard CB et al. Evidence for a role of glucose-induced translocation of glucokinase in the control of hepatic glycogen synthesis. J Biol Chem 1996; 271:30479-30486.
31. Fajans SS. Maturity onset diabetes of the young (MODY). Diabetes Metab Rev 1989; 5:579-606.
32. Nadeau JH. Modifying the message. Science 2003; 301:927-928.
33. Ideker T, Thorsson V, Ranish et al. Integrated genomic and proteomic analyses of a systematically perturbed metabolic network. Science 2001; 292:929-934.
34. Cornish-Bowden A, Cárdenas ML. Complex networks of interactions connect genes to phenotypes. Trends Biochem Sci 2001; 26:463-465.
35. Sved JA, Mayo O. The evolution of dominance in Mathematical Topics in Population Genetics In : K-I Kojima, ed. Springer, Berlin: 1970 :289-316.
36. Ohh BK, Sheldon BL. Selection for dominance of hairy-wing (Hw) in Drosophila melanogaster. I. Dominance at different levels of phenotype. Genetics 1970; 66:517-540.
37. Mayo O, Bürger R. Evolution of dominance: A theory whose time has passed? Biol Rev 1997; 72:97-110.
38. Doebley J, Stec A, Gustus C. Teosinte branched1 and the origin of maize: Evidence for epistasis and the evolution of dominance. Genetics 1995; 141:333-346.
39. Otto SP, Bourguet D. Balanced polymorphisms and the evolution of dominance. Amer Nat 1999; 153:561-574.

40. Chandra HS, Nanjundiah V. Dosage compensation from the viewpoint of natural selection. In: Reed KC, Graves JAM, eds. Sex Chromosomes and Sex-determining Genes. Reading: Harwood Academic, 1993:215-242.
41. Cook DL, Gerber AN, Tapscott SJ. Modeling stochastic gene expression: Implications for haploinsufficiency. Proc Natl Acad Sci USA 1998; 95:15641-15646 .
42. Veitia RA. Exploring the etiology of haploinsufficiency. BioEssays 2002; 24:175-184.
43. Hurst LD. Evolutionary genomics: Sex and the X. Nature 2001; 411:149-150.
44. Omholt SW, Plahte E, Øyehaug L et al. Gene regulatory networks generating the phenomenona of additivity, dominance and epistasis. Genetics 2000; 155:969-980.
45. Papp B, Pál C, Hurst LD. Dosage sensitivity and the evolution of gene families in yeast. Nature 2003; 424:194-197.
46. Waddington CH. In: The Strategy of the Genes. London: George Allen and Unwin, 1957.
47. Debat V, David P. Mapping phenotypes: Canalization, plasticity and developmental stability. Trends Ecol Evol 2001; 16:555-561.
48. Hartman JL IV, Garvik B, Hartwell L. Principles for the buffering of genetic variation. Science 2001; 291:1001-1004.
49. Forsdyke DR. The heat-shock response and the molecular basis of genetic dominance. J Theor Biol 1994; 167:1-5.
50. Gilchrist MA, Nijhout HF. Nonlinear developmental processes as sources of dominance. Genetics 2001; 159:423-432.
51. Zeyl C, Vanderford T, Carter M. An evolutionary advantage of haploidy in large yeast populations. Science 2003; 299:555-559.

Chapter 2

Phenomenology and Mechanistics of Dominance

Reiner Veitia and Bruno Bost

Abstract

Here we explore the phenomenological and mechanistic bases of dominance. Dominance can be defined as an operational measure of the deviation of the heterozygote's observed trait value from its expected value based on the phenotype of the two homozygotes. A simple mathematical analysis of various measures of dominance shows that its main source is the existence of nonlinear relationships between the genotypic and phenotypic values. However, the mechanisms generating these nonlinearities are diverse. We will outline the molecular mechanisms of various cases of dominant phenotypes. As dosage imbalances lead to an important class of dominant mutations we attempt to explain these effects from a theoretical perspective. Finally, we discuss issues concerning robustness against deleterious mutations from the perspective of gene redundancy, the topology of cellular networks and other buffering effects to maintain a normal phenotype .

Introduction

Mendel[1] talked about dominance when analyzing characters of cross-pollinating true-breeding garden peas over 150 years ago. In modern terms, when a homozygous individual AA is crossed with a homozygous individual aa, the offspring (F1) will be heterozygous Aa. If the first generation F1 individuals have the same characteristics as one of the homozygous individuals, this character is said to be dominant. The character that disappeared from the F1 is called the recessive one. It reappears in 25% of the offspring after the cross of two F1 individuals (25% AA, 50% Aa and 25% aa). It must be noted that, in principle, dominance and recessivity refer to characters, the phenotypic result of gene action not to genes or alleles (variants of a gene). Dominance is a shorthand term to describe the relationship of the phenotypes of three or more genotypes.[2] However, "dominant" is currently applied to characters or to alleles, a dual usage that may lead to some confusion. In terms of characters, a dominant trait masks or obscures the expression of an alternative form of that character. This loose definition does not even require the alternative form to be allelic (i.e., arising from the same locus). In contrast, defining a dominant allele does require that the alternative form be controlled by another allele (at the same locus). Accordingly, a dominant allele controls a character which masks or obscures the expression of an alternative form controlled by the recessive allele.[3,4]

Dominance and recessivity, as used in introductory genetic courses, are rather caricatural terms because a particular genotype does not always determine one distinct phenotype. These terms were coined after the study of "qualitative" characters where only two states could be scored. In many simple instances, several alleles can be alternatives for a single locus, giving rise to allelic series. The alleles of the series can generate phenotypes whose degrees of dominance can be different. This is the case, for example, of albino mutations in rodents, where deleterious alleles exhibit additive relationships whereas they are all recessive when compared with the wild-type allele.[5,6] More in depth quantitative assessments of the phenotypic effects of a mutation led Müller in 1932[7] to propose a classification of dominant mutations. On purely genetic grounds he proposed the terms amorph, hypomorph and hypermorph (among others) to reflect gross quantitative changes with respect to a wild-type character. An amorph allele is what we currently call a null allele, that is, the one whose product is non functional. A hypomorph is a non null allele which generates a phenotypic value lower than that of the wild-type, whereas a hypermorph leads to phenotypic values higher than the one of the wild-type.[7] With the development of genetics, more quantitative notions to assess dominance were required, and will be dealt with in a specific section.

A satisfactory mechanistic theory of dominance should be able to explain why recessive mutations are more common than dominant ones[8-11] and linked to this, why the heterozygote presents most often the same phenotype as one of the homozygotes. It should also explain why strongly deleterious mutations are mostly recessive; and why mutations with small effect display additivity but in a deleterious allelic series, mutant alleles are additive among them but are recessive with respect to the wild-type.

The reader must be aware that, as pointed out by Miller,[3,4] epistasis, which refers to the interaction between loci can give rise to apparent cases of dominance. Thus, a character can be "dominant" while in genetic terms it arises from the epistatic interaction between two loci. For instance, alternative forms such as black and white plumage color in a particular stock of chickens can be controlled by nonalleles from interacting loci (nonallelic). The character black seems to be dominant to the character white as the cross of purebred parental types yields 75% black and 25% white progeny. However, a cross between the white stock and the wild-type produces the typical phenotypic ratios of a dihybrid cross (i.e., two loci are considered) that includes a recessive epistatic gene.[3]

In this chapter, we will explore different phenomenological and mechanistic explanations of dominance. We play with concepts with the sole purpose to illustrate how complex matters are. However, we do not attempt to provide a definitive answer to the problems we will discuss. Hopefully, this overview will allow the readers to formulate their own opinions.

Nonlinearity and Dominance

Dominance can be defined as an operational measure of the deviation of the heterozygote's observed trait value from its expected value based on the phenotype of the two homozygotes.[12] This operational definition presumes that we are able to measure the phenotypic outcome of a certain genotype.

An early measure of dominance was introduced by Sewall Wright.[13] His dominance index was defined as:

$$D = (P_W - P_H) / (P_W - P_M)$$

where P_W, P_H and P_M are the phenotypic values of the wild-type homozygote (W), the heterozygote (H) and the mutant homozygote (M), respectively. It is clear that the scale of D runs from 0 when $P_H = P_W$ (M is recessive), through $D = 0.5$ (intermediacy or "additivity"), to $D=1$ when $P_H = P_M$ (W is recessive).

Another commonly used measure of dominance is Falconer's degree of dominance (Falconer and Mackay[14]):

$$d = (P_H - (P_W + P_M) / 2) / ((P_W - P_M) / 2)$$

The scale of d runs from -1 when $P_H = P_M$ (W is recessive), through d = 0 (additivity), to d=1 when $P_H = P_W$ (M is recessive). It is easy to show that these two indexes are negatively related by:

$$d = 1 - 2D$$

A superficial mathematical analysis of these measures leads to the conclusion that the existence of nonlinear relationships between the genotypic and phenotypic values leads to dominance. As an example, we can consider two alleles A (wild-type) and a (mutant), whose allelic values are respectively g_A and g_a. The phenotypic value P_i of a diploid individual is a function of its genotypic value G_i, with $G_i = \vartheta_A g_A + \vartheta_a g_a$, where ϑ_A and ϑ_a are respectively the number of alleles A and a (for the heterozygote, $\vartheta_A = \vartheta_a = 1$). The genotypic value is then determined in a strictly additive way. In the case of a linear relationship between the genotypic and phenotypic values, the dominance indexes are D = 0.5 and d = 0, showing additivity. Dominance occurs when the phenotypic value of the heterozygote is not halfway between the phenotypic values of the homozygotes. Dominance relationships are very different when nonlinear relationships follow, for instance, Hill's equation: $P_i = G_i^{\,n} / (K + G_i^{\,n})$. With K = 0.5 and n = 1, we get a hyperbola, typical of metabolic pathways when considering the relationship between flux and enzyme concentration (Fig. 1A). With K = 1 and n = 5, we get a sigmoid, which can be found in certain metabolic pathways or in transcriptional response (Fig. 1B). Varying the genetic values g_A and g_a, with those two nonlinear functions (i.e., n=1 or 5 in Hill equation), we thus get very different dominance indexes (Table 1). In metabolic systems, the dominance index (Wright's one) runs from values close to 0.5 (additivity) in linear parts of the hyperbola (i.e., g_A = 0.25 and g_a = 0.05 for instance) to values close to 0 (mutation is strongly recessive) when heterozygote is on the plateau (i.e., g_A = 5.00 and g_a = 0.00). With a sigmoidal relationship, the dominance index depends on the position of the individuals compared to the inflexion point of the curve. When the heterozygote is near this point (i.e., g_A = 1.00 and g_a = 0.00), we have additivity (D ~ 0.5). In this region, small variations of G will lead to dramatic variations in P. By contrast, when the heterozygote deviates from the inflexion point toward one homozygous phenotype, we get strong dominance of this phenotype. Thus mutation can run from

Table 1. Effects of non-linear relationships between genotypic (G_i) and phenotypic (P_i) values on dominance indexes

Genetic Values		Hyperbolic Relationship K = 0.5 / n = 1		Sigmoidal Relationship K = 1 / n = 5	
g_A	g_a	D	d	D	d
0.25	0.05	0.38	0.25	0.92	-0.84
1.00	0.00	0.17	0.67	0.48	0.03
5.00	0.00	0.05	0.90	0.00	1.00
5.00	4.00	0.45	0.11	0.34	0.32

D: Wright's index, d: Falconer's index), in diploid individuals, considering two alleles A (wild-type) and a (mutant), which allelic values are respectively g_A and g_a. The relationship between P_i and G_i follows Hill's equation: $P_i = G_i^{\,n} / (K + G_i^{\,n})$ and G_i is determined in a strictly additive way from the allelic values $G_i = \vartheta_A g_A + \vartheta_a g_a$, where ϑ_A and ϑ_a are the number of alleles A and a, respectively. As a reference, when the relationship between genotypic and phenotypic values is linear, D = 0.5 and d = 0 (additivity). Parameters of the equations are arbitrary and are only intended to illustrate concepts.

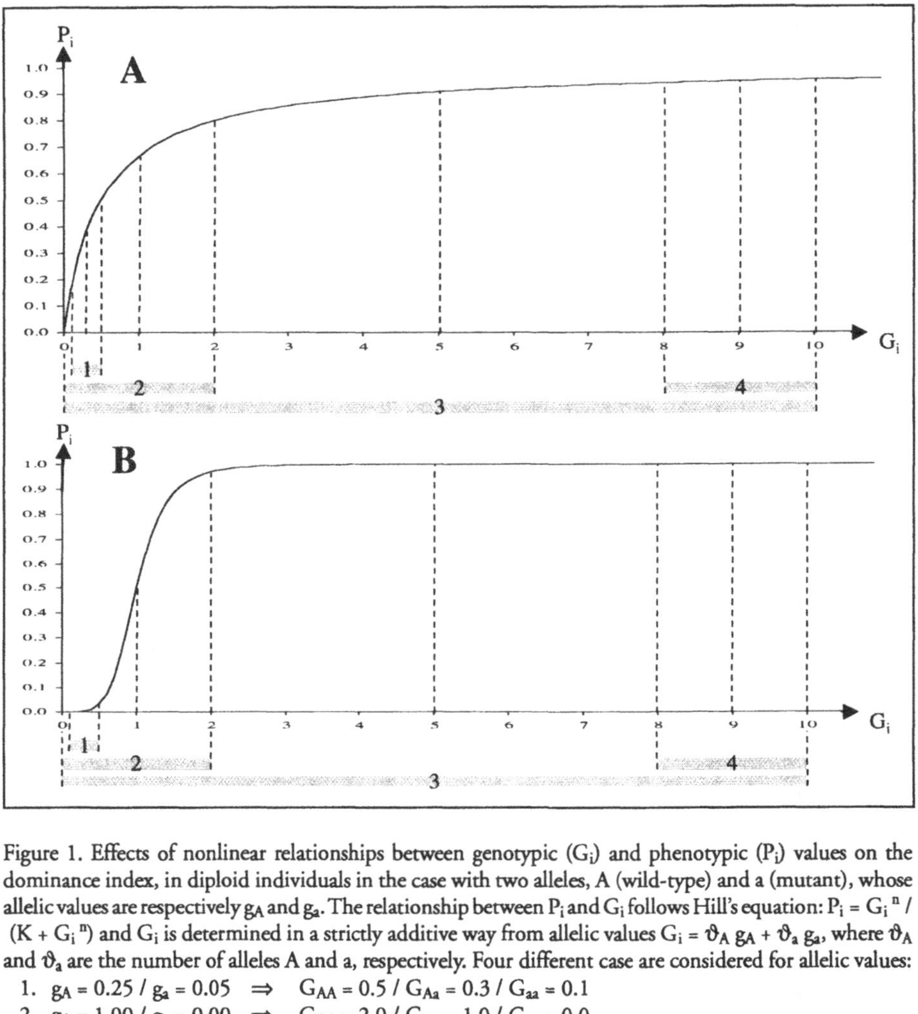

Figure 1. Effects of nonlinear relationships between genotypic (G_i) and phenotypic (P_i) values on the dominance index, in diploid individuals in the case with two alleles, A (wild-type) and a (mutant), whose allelic values are respectively g_A and g_a. The relationship between P_i and G_i follows Hill's equation: $P_i = G_i^{\,n} / (K + G_i^{\,n})$ and G_i is determined in a strictly additive way from allelic values $G_i = \vartheta_A\, g_A + \vartheta_a\, g_a$, where ϑ_A and ϑ_a are the number of alleles A and a, respectively. Four different case are considered for allelic values:

1. $g_A = 0.25 / g_a = 0.05 \Rightarrow G_{AA} = 0.5 / G_{Aa} = 0.3 / G_{aa} = 0.1$
2. $g_A = 1.00 / g_a = 0.00 \Rightarrow G_{AA} = 2.0 / G_{Aa} = 1.0 / G_{aa} = 0.0$
3. $g_A = 5.00 / g_a = 0.00 \Rightarrow G_{AA} = 10.0 / G_{Aa} = 5.0 / G_{aa} = 0.0$
4. $g_A = 5.00 / g_a = 4.00 \Rightarrow G_{AA} = 10.0 / G_{Aa} = 9.0 / G_{aa} = 8.0$

Two different relationships are considered, varying Hill's coefficient:
Panel A) Hyperbolic relationship between genotypic and phenotypic values: $P_i = G_i / (0.5 + G_i)$
Panel B) Sigmoidal relationship between genotypic and phenotypic values: $P_i = G_i^{\,5} / (1 + G_i^{\,5})$
In all cases, when Gi elicits a P around the plateau of the curves, the normal phenotype is dominant.

quasi-complete recessivity (i.e., $g_A = 5.00$ and $g_a = 0.00$) to quasi-complete dominance ($g_A = 0.25$ and $g_a = 0.15$).

The effect of nonlinearities in the determination of complex phenotypes (i.e., quantitative traits), in terms of decomposition of genetic variance in populations, has been studied theoretically with various models.[12,15-18] These studies show that these nonlinearities in complex traits determination generate dominance, even with additive behavior at the genetic level.

The fact that nonlinear relationships between genotypic and phenotypic values lead to dominance, was noted very early by Wright.[13] Perhaps on the basis of the nascent theory of enzyme kinetics he "guessed" that the reaction flux in a catalyzed pathway could follow the law of diminishing returns (hyperbola) as a function of enzyme concentration. If so, when enzyme activity levels (E) are near the asymptote of the curve, changes in activity have very little effect on flux (Fig. 1A). This physiological explanation for dominance relationships was in contradiction with Fisher's theory, where dominance was supposed to arise from the selection of alleles at "modifier" loci that change the heterozygote phenotype making it equivalent to the wild-type.[8,19,20] The controversy went on for several years, until Crosby[21] and Charlesworth[22] demonstrated that Fisher's hypothesis was unlikely to be correct. The strongest argument against the existence of selection for dominance was given by the work of Orr[23] on the haplodiplontic alga *Chlamydomonas reinhardtii.* He found that most mutations, particularly lethal mutations, are recessive. As the diploid phase of the alga is rare and brief, and as the lethal mutations are eliminated in the haploid phase, the intervention of modifiers giving a selective advantage to the heterozygotes is very unlikely and suggests that dominance relationships come from the structure and functionning of the organism.

A theory of dominance from the point of view of the analysis of changes in enzyme activity (concentration or specific activity) was put forward by Kacser and Burns in 1973.[24] It is remarkable how genetics waited for more than a century since Mendel formulated the laws of inheritance to find a successful, yet simple, explanation of dominance. In papers, which are now considered to be classics, Kacser and Burns[2,24] developed a modern theory of the steady state of metabolic pathways known as Metabolic Control Analysis (MCA). They defined what is today called the flux control coefficient C as the rate of change of the flux J following an infinitesimal change in the concentration of the enzyme. They and other groups[25] showed that the sum of control coefficients of all the enzymes involved in a particular flux =1. From this, it follows that given the large numbers of enzymes involved in a pathway, in general C values are small. This immediately explains the well known fact that diminution in the amount (activity) of the enzyme by as much as 50% (in a heterozygote) does not affect visibly the phenotype (flux) and the normal phenotype is dominant. This was considered by the authors as an inevitable kinetic behavior and as evidence that selection was not necessary to shape such types of systems.

Further analyses have shown that there are situations in which moderate changes in enzyme concentration can induce substantial changes in flux. For instance, this may be the case of hypothetical pathways operating in conditions where all the enzymes are more than half saturated[26] or in the presence of significant nonlinearities such as for enzymes with Hill coefficients ranging between two and six.[27] From this, it seems that dominance is not an inevitable[26] trivial "default" consequence of physiology[27] and that selection may have acted to avoid systems in which several steps have high C values. In line with this, recent results show that for mutations involving finite changes (of any magnitude) in enzyme concentration, the flux summation theorem outlined above only holds in a very restricted set of conditions. In fact, it seems to be valid only in cases where the relationship between genotype and phenotype is linear and devoid of nonlinearities in the form of epistasis. An absence of epistasis is unlikely in metabolic systems as it can arise from nonlinearities such as those caused by enzyme saturation. In such cases dominance levels can be modified by mutations that affect saturation levels. It follows then, that dominance is not a necessary property of metabolic systems and that it can be subject to evolutionary modification.[28]

In spite of these theoretical criticisms, the MCA remains the simplest and most elegant way to explain why most mutations in enzymes are recessive (see Porteous 1996[29] for further discussion on MCA). A recent analysis of human disease genes has clearly confirmed these theoretical predictions.[30] In an extensive search of databases such as the Metabolic and Molecular Bases of Inherited Disease and the Online Mendelian Inheritance of Man, the authors detected 923 disease genes. The category including enzymes accounts for 31.2% of this figure and for more than 75% of these enzymes, disease-causing mutations were recessive, as expected on theoretical grounds.[31]

Today it is widely admitted that dominance is a consequence of nonlinear relationships between genotype and phenotype due to the structure and behavior of biological systems, rather than a consequence of a specific selection as supposed by Fisher. However, as outlined above evolutionary pressures can influence dominance relationships. In case of metabolic pathways, as control is distributed among all the enzymes, it is easy to show that modification of an enzyme-encoding gene affects the dominance relationships of all the other enzymes in the pathway. This effect has been studied in theoretical pathways.[32] Thus, genes involved in a metabolic pathway can be viewed as epistatic "modifiers" of dominance of other genes in the same pathway. More generally, the selection of modifiers acting on dominance relationships has been studied, in order to predict how dominance can evolve, rather than to explain the origin of dominance.[33-35] It is likely that a great number of genes can modify dominance relationships of other genes involved in the determination of the same trait. Nonlinear behavior and interactions between genes are therefore a fertile ground for generating dominance and epistasis.

Penetrance and Expressivity: The Quantitative Dimension of Dominance and Epistasis

Penetrance is a statistical notion that refers to the proportion of individuals that display a given phenotype supposed to reflect a particular genotype. This implies that: either i) we are unable to detect the phenotypic manifestation of a mutation with absolute certainty and/or ii) the phenotypic effects of certain mutations can be silenced or buffered by the "genetic background". In fact, both cases are likely to be true. It is known that penetrance depends on the method used to assess the phenotype. The closer one is to gene action, the higher the observed penetrance will be. When one focuses on all heterozygous individuals expressing the phenotype expected for the mutation, the degree of phenotypic manifestation of the mutation can be different from one individual to another. This phenomenon is called "variable expressivity". Variable penetrance and expressivity are shorthand terms that reflect our ignorance concerning the molecular mechanisms and the genetics of mutations that lead to dominant phenotypes (see Fig. 2). Glimpses of the mechanistic basis of penetrance and expressivity are already at hand. As an example we have the inheritance patterns seen in some complex traits, such as type I diabetes, where a dominant HLA genotype is necessary, but not sufficient, to cause disease. A second, recessive alteration, namely, a promoter polymorphism of the insulin gene, must also be present.[36] A model of a dominant loss-of-function mutation as a prerequisite to unmask recessive mutations may explain the genetic heterogeneity seen in inbred populations, where both one dominant and two recessive mutations are required for pathogenesis. The presence of a dominant "master locus" interacting with different "partners" may also explain the difficulty in replicating linkage data from discrete populations for various complex traits such as asthma, diabetes, and schizophrenia, because each population isolate will be enriched for a limited number of the recessive mutations that alone cannot generate pathogenicity. These types of inter-loci interactions may lead to transmission models that bridges the gap between Mendelian disorders and complex traits.[37]

The Molecular Biology of Dominance

It is difficult to conceive a single theoretical model of dominance able to account for the many dissimilar observations described in the literature so far. For instance, the MCA provides satisfying explanations for the dominance of wild-type phenotypes but is not convincing when dealing with dominant mutations. Furthermore, the MCA is not expected to provide explanations for dominance in the context of macromolecular complexes, within signaling pathways or transcription networks. The heterogeneous nature of the sources of dominance

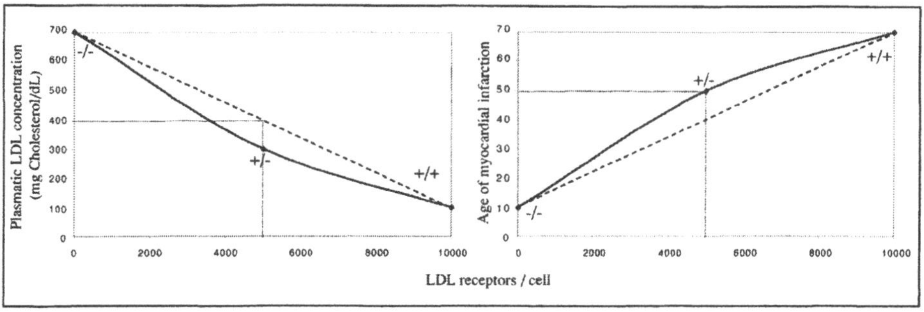

Figure 2. Nonlinearities in familial hypercholesterolemia (FHC) in man, a disease that can be due to mutations in the receptor of low density lipoproteins (LDLR). The plasma LDL concentration (left panel) and the mean age of myocardial infarction (right panel) are plotted against the number of LDLR per cell. Symbols +/+, +/- and -/- stand for the normal, heterozygote and double mutant respectively. Due to nonlinearities, halving the density of receptors (in the heterozygote +/-) the plasma concentration of "bad" cholesterol is lower than expected if the relationship was linear. Fortunately, this translates into a delay of about 10 years of the heart attack (with respect to a linear relationship). This also shows the relativity of the term "dominance". With direct DNA/protein analysis the alleles are codominant, they can both be expressed. When one assesses the plasmatic LDL-cholesterol, the normal phenotype (and allele) displays some degree of dominance. However, concerning myocardial infarction over the lifetime FHC (abnormal phenotype) is considered to be dominant. FHC leads to xanthomatosis (lipid storage) (see ref. 103). In a consanguineous kindred containing several individuals with homozygous LDLR mutations, half of them had giant xanthomas, while the other did not, even when their LDL-cholesterol concentrations were elevated to similar degrees. Heterozygous FHC individuals were also clearly distinguishable with respect to xanthoma size. Segregation analysis suggested that a separate susceptibility gene may explain the formation of giant xanthomas.[104] Furthermore, the two heterozygous sibs of a girl with severe HC (compound heterozygote for LDLR mutations) carried one of her mutations, but only one of them was hypercholesterolemic. Accordingly, there may be cholesterol-lowering mechanisms depending on modifier genes.[105]

is also apparent from the study of its possible molecular mechanisms. Without being exhaustive, the main known mechanisms generating dominance can be outlined as follows:

- Reduced gene dosage, expression or protein activty: haploinsufficiency,
- Increased gene expression and increased or constitutive protein activity,
- Dominant negative effects (antimorphs), altered structural proteins, abnormal subunit aggregation and trans-dominance,
- Ectopic or temporally altered mRNA or protein expression,
- Toxic deposition of abnormal proteins,
- New protein functions/chimeric proteins,
- Apparent dominant phenotype with recessivity at the cellular level. Dominantly inherited cancers,
- Non allelic non complementation,
- and Polar overdominance.

The reader will notice that this is a rough classification and that different molecular mechanisms may coexist within each group. For complementary informations see ref. 38

Haploinsufficiency: Reduced Gene Dosage, Expression or Protein Activity

There are situations where dominance arises from the total, or partial, lack of activity of one allele at a diploid locus, that is, half of the normal amount of the gene product is insufficient to maintain a normal phenotype. This phenomenon is called haploinsufficiency (HI). Loss-of-function mutations leading to HI may be the result of heterozygous deletions, promoter or splice site mutations, or intragenic mutations leading, in the most severe cases, to half the normal amount of active product with respect to the wild type homozygote. It is worth noting that HI, although deleterious, is compatible with survival of the individual. However, there are many instances in which absence of one allele at a locus will induce death at early stages of development leading to what is known as haplolethality (HL). HL can be considered as the extreme of HI. The clearest examples of HI come from cases where there is a heterozygous deletion of a locus. The analysis of a series of HI proteins that we will describe below shows that, in significant number of cases, these factors are involved in the formation of multimeric structures. Other factors are involved in cascades of signal transduction or transcriptional control, where many interactions are required to pass the information from the input signal to the biological output. In most cases, steps leading to nonlinear amplification of the signal are involved.

Wilkie[38] has broadly divided haploinsufficient loci into two main groups: those producing proteins synthesized and required in large amounts and those coding for regulatory proteins that normally work close to a threshold (i.e., transcription factors). A plausible way to account for the behavior of seemingly unrelated gene groups is to suppose that the relationship connecting their genotypic and phenotypic values is nonlinear, but not hyperbolic as proposed by Wright or Kacser and Burns, but rather sigmoidal (Fig. 1B). At least in some cases of HI, the deletion of one allele drives the genotypic value (i.e., transcription factor concentration) near to the inflexion point of the sigmoid. In such cases, the phenotypic outcome depends dramatically on the particular genotypic value, the action of modifiers and the environment.

In line with the ideas discussed above, HI has been well established in man for the Collagens (IIA1 and VA1) that participate in the formation of connective tissue.[39,40] It has been shown that collagen fibril assembly involves initial formation of micro-fibrils through a helical cooperative mechanism (a very nonlinear process). Thus, at least in vitro, type I collagen self-associates efficiently only if the concentration of the protein is above a critical value.[41] HI has also been well established in man for elastin, another polymeric component of the connective tissue.[42] Furthermore, it has recently been shown that its coacervation (polymerization) is a highly cooperative monomer→oligomer self-association reaction requiring a critical monomer concentration.[43] Again in agreement with the notion of nonlinearity, HI of protamines 1 or 2 is known to cause infertility in mice. Protamines are the major DNA-binding proteins in the nucleus of sperm in most vertebrates and package the DNA into a volume less than 5% of a somatic cell nucleus. A decrease in the amount of either protamine disrupts sperm nuclear formation and normal sperm function.[44] As expected, protamine-DNA interactions display an unusually high degree of cooperativity.[45]

A myriad of examples of HI transcription factors have been documented in man.[46,47] As discussed in these references, the nonlinear response of transcription to transcription factor concentration, due to cooperative promoter recognition and synergistic effects may be responsible for HI. This response depends on many parameters which opens the possibility to the existence of compensatory mutations and modifiers of dominance.[48] Stochastic gene expression has also been incriminated as a possible source of HI. Indeed, simple models of stochastic gene expression have shown that diploid systems have a higher probability of uninterrupted gene expression and more predictable responses than a corresponding haploid state. HI would result from an increased susceptibility to stochastic delays of gene expression initiation or interruptions of expression.[49]

Within the context of HI and dominance, it is relevant to consider the case of monoallelic expression. This has been linked for long time to the phenomenon of parental genomic imprinting. For a certain set of genes (imprinted) only one of the alleles (maternal or paternal) is transcribed. The imprinting status of a gene may depend on the tissue and developmental stage. In the case of genes undergoing tissue/stage-specific imprinting, HI can lead to complex patterns of inherited phenotypes. One phenotype would result from plain HI of the gene

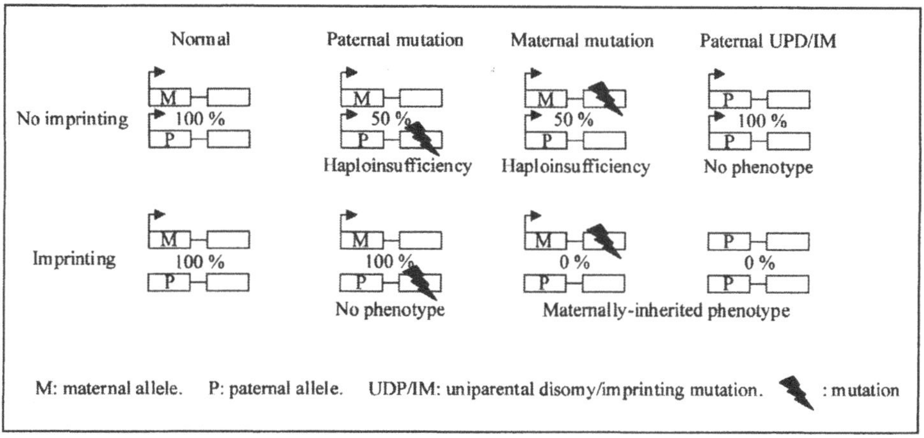

Figure 3. Effect of parental imprinting on the pattern of inheritance. Upper row: The gene is bi-allelically expressed (as most genes). A deletion or mutation inactivating one allele will lead to a phenotype if the mutation is "dominant" (i.e., haploinsufficiency). Lower row: If only the maternal allele is expressed (imprinting), a mutation inactivating the paternal allele has no effect. If the mutation inactivates the maternal allele (for instance an imprinting mutation, IM) or in case of uniparental disomy UPD (i.e., two copies of the paternal locus in the example), an abnormal phenotype appears due to complete lack of gene product. In this particular case, the mode of inheritance of the mutation is non Mendelian and is transmitted only by the mother. Modified from reference 50.

product in tissues where the gene is not imprinted and can be inherited from either parent. An other phenotype would be caused by the total loss of the gene product in tissues where the gene is imprinted and this would only be manifested when the defect affects the parental allele which should be active.[50] As an example, consider the gene GNAS1 which codes for a stimulatory GTP binding protein Gs that stimulates adenylate cyclase. GNAS1 undergoes genomic imprinting resulting in the paternal allele being silenced in a tissue-specific manner (for instance, in the renal tubules). Mutations in the maternal allele lead to Albright hereditary osteodystrophy (AHO), accompanied by an end-organ resistance to several hormones that activate Gs-coupled pathways. This translates mainly into pseudohypoparathyrodism. In contrast, paternal mutations lead to isolated AHO. The AHO phenotype may be the result of plain Gs(α) HI whereas multihormone resistance in patients carrying maternal but not paternal mutations can be explained by tissue-specific imprinting of the gene, with silencing of the paternal allele in many hormone target tissues[50] (Fig. 3, modified from ref. 50).

Monoallelic expression of diploid genes may also result from a random silencing of one of the alleles of a given locus.[51] In these cases, tissues are a mosaic with respect to the transcribed allele. This has been shown for olfactory receptors, some interleukins and the gene Pax5. Inactivation of one allele will either have no phenotypic consequences in the cells where this allele is not expressed or cause a total lack of the active gene product if the expressed allele bears the mutation. This seems to be the specific explanation for PAX5 HI. Nutt and Bussilinger[51] have proposed that this phenomenon could underlie HI due to mutations in other PAX genes (like PAX1, 2, 3, 6 and 8). It remains to be shown whether this is the case or not for these particular examples. However, it is conceivable that many more genes exist where monoallelic expression underlies HI.

The MCA theory successfully predicts that HI should not be observed for enzymes. However, there is an outstanding case in humans, concerning the enzymes involved in heme/porphyrin biosynthesis. The defects of at least five enzymes lead to various types of dominant porphyria showing that flux of the pathway is tightly dependent on the concentration of many enzymes. Symptoms are associated with increased activity of the delta-aminolevulinate synthetase which is the first and most controlling step of the pathway (i.e., the enzyme with the highest control coefficient). This enzyme undergoes negative feedback by the heme group, the end product. Decreased heme concentrations due to defects in downstream enzymes are thought to be responsible for derepression of the delta-aminolevulinate synthase which explains the increased activity detected in sick individuals. The environment, including chemicals that stimulate enzyme production play a role in the development of clinical symptoms which are relieved with the supply of exogenous heme.[52]

Increased Gene Dosage and Increased or Constitutive Protein Activity

In this section we can conceive several cases: (i) the mutant protein is intrinsically more active than the wild-type, (ii) the protein activity can not be turned off, (iii) the protein activity is expressed at a higher level due to incorrect transcriptional or translational control, or (iv) increased RNA or protein stability. Evidence from trisomies in humans and other organisms suggests that increased gene dosage leads to dominant phenotypes. Molecular details are available for some genes whose increased dosage results in a gain-of-function that produces new phenotypes. This is the case of PLP and PMP22, two genes involved in myelin formation. One encodes the proteolipid protein PLP of the central nervous system and the other encodes the peripheral myelin protein 22 (PMP-22). Their duplications lead to Pelizaeus-Merzbacher disease/spastic paraplegia type II and the type 1A Charcot-Marie Tooth syndrome respectively.[53]

The SOX9 gene (SRY-related HMG box 9) is another interesting example. Its HI is responsible for bone anomalies and gonadal dysgenesis in a 46,XY background.[54] However, a duplication of a genomic region containing SOX9 is associated with female to male sex reversal. Over-expression of SOX9 seems to be the most likely explanation for the phenotype, which is testis formation in the absence of SRY, the master testis determining gene.[55] The mouse Odsex mutation (Ods) adds credence to this notion. In the XX Ods/+ mouse, a transgene inserted in a putative regulatory region of Sox9 can de-repress its expression producing a male phenotype in an XX background.[56]

Another example of increased gene dosage leading to abnormal function comes from the constitutive over-expression of Id1, an inhibitor of the DNA binding capacity of bHLH (helix-loop-helix) proteins. This leads to a phenotype that resembles that of the null mutation for E2A, a factor involved in B cell development.[57] In this case, titration of bHLH proteins, playing pivotal roles in B cell development, by the inhibitor Idl is responsible for the phenotype. There are many other examples of increased or constitutive protein activity causing dominance and we will mention three of these. In a rare case of gout, an increased level of phosphoribosyl-pyrophosphate synthase, due to a structural lesion of this enzyme, leads to a higher specific enzymatic activity.[58] Mutations in the proto-oncogene Ras that prevent GTP hydrolysis maintain the protein in an activated state and render it independent of the inductive signal.[59,60] Finally, activating mutations in the GTP binding protein Gs mentioned above, are responsible for the McCune-Albright syndrome, which is characterized by polyostotic fibrous dysplasia, cafe au lait spots, sexual precocity, and hyperfunction of multiple endocrine glands.[60]

Dominant Negative Effects, Altered Structural Proteins and Abnormal Subunit Aggregation

There are mutations, which do not lead to a complete absence of transcription or translation of one allele from the locus, but rather induce dominant phenotypes through a different mechanism. Here, the production of a mutant protein (from one allele) interferes with the action of the normal one (from the other allele). This phenomenon is called a dominant negative effect and the alleles producing it are called antimorphs.[61] In the case of missense mutations, it is particularly difficult to determine a priori whether HI or a dominant negative effect is responsible for the phenotypes. In addition, there is an increasing number of examples where different mutations in the same gene lead either to HI or induce dominant negative effects. The most intuitively clear examples of dominant negative effects come from factors that oligomerize. In these instances, the mutant polypeptide will poison the complex. For example, if the protein acts as a homodimer, the dimer may be unable to properly function if it contains a normal subunit A, sequestered by the nonfunctional one A'. In the case of a homodimer one will get three inactive molecules (two AA' and one A'A') per active molecule AA.

From the above discussion it is clear that the same gene can lead to HI while some of its mutations can lead to a dominant negative effect. One such example is the transcription factor WT-1, which exhibits HI in human gonadal formation. However, one mechanism recognized as being involved in Wilms' tumor (nephroblastoma) formation, implies the existence of dominant negative mutations. WT-1 is known to interact with other transcription factors, such as p53, which also display dominant negative mutations.[62,63]

There are other cases where the dominant negative effects arise from interactions between different proteins. For instance, the E1A protein of adenovirus stimulates the transcription of early viral and certain cellular genes. Trans-activation by E1A is mediated through cellular transcription factors. The ability of the E1A point mutant (S185N) to inhibit wild-type E1A trans-activation is thought to result from the sequestration of a limiting cellular transcription factor. This is a case of trans-dominance: the mutant is a trans-dominant negative.[64] Similarly, the HTLV-1 Rex protein is an essential shuttle required for nuclear export of unspliced and incompletely-spliced viral RNAs. A naturally occurring Rex mutant lacking the RNA binding domain is highly trans-dominant. It has been proposed that trans-dominant Rex mutants do not function by retaining WT Rex in the nucleus by protein-protein interactions but rather titrate factors essential for Rex/Rev export.[65]

Mutations of structural proteins that assemble into polymers are frequently dominant. A heterozygous individual has a mixture of normal and abnormal protein molecules that interfere with proper formation of the aggregated proteins. For instance, mutant molecules of type I collagen can poison collagen fibrils and lead to *osteogenesis imperfecta*. The extent of fibril disruption correlates, as expected, with the severity of the phenotype. A similar mechanism explains Marfan syndrome, which results from mutations in fibrillin. Several dysfibirinogenemias (a tendency to bleed) result from an admixture of normal and mutant fibrinogen molecules. The boundary between this type of mutation and dominant negative effects are so unclear that in some references they are considered as a single entity.[66] That is why we have considered them together. A similar mechanism is obvious in keratin disorders. Keratins are intermediate filament proteins that form a cytoskeletal network within epithelial cells. Dominant mutations in the genes encoding these proteins disrupt the keratin filament cytoskeleton resulting in cells that are less resilient and blister after mild physical trauma. Mutations have now been identified in 18 keratins, all of which produce a fragile cell phenotype and lead to disease (i.e., epidermolysis bullosa simplex, ichthyosis bullosa of Siemens, epidermolytic palmoplantar keratoderma, pachyonychia congenita, white sponge nevus, Meesmann's corneal dystrophy, cryptogenic cirrhosis and monilethrix[67]).

Ectopic or Temporarily Altered mRNA Expression

Many examples, including natural, induced or transgenic mutants can be cited in this section. A striking case of loss of temporal regulation is the hereditary persistence of fetal hemoglobin in man. A mutation in the promoter of the gamma globin [68] and deletions encompassing the delta and beta globin genes prevent the switch from expression of the fetal gamma globin to the delta and beta globins that operates around birth.[69]

Transgenic mice, where heterologous promoters drive transcription of a given Hox gene in the head provide a further example. These studies have shown that the more posteriorly expressed Hox genes (in natural conditions) tend to have a significant effect only on the skull bones of mesodermal origin whereas those normally expressed more anteriorly, in the hindbrain and branchial arches, can affect more anterior branchial arch and neural crest-derived structures. Manipulations triggering homeobox expression in small, localized regions of the facial precursors show the dramatic effects of this altered expression on facial development. Retinoic acid, a well known morphogen (and teratogen), induces Hox gene expression and at least some of its teratogenic action may be mediated by its up-regulation of the Hox and other homeobox genes in the head.[70]

Toxic Deposition of Abnormal Proteins

The best characterized examples in this group are the amyloidoses. This heterogeneous class includes diseases associated with the deposition of insoluble aggregates of mutant proteins (that are normally soluble). The best known proteins are transthyretin (prealbumin), gelsolin, fibrinogen, lysozyme, Ig light chains, amyloid A, the prion protein and beta 2-microglobulin. In all, a total of 18 proteins have been definitively identified as amyloid precursors associated with human disease. Mutations in the genes that encode some of these proteins produce autosomal dominant disease in mid to late adult life.[71] A very interesting set of diseases that can be included in this group are those where a dynamic mutation leads to the expansion of an aminoacid repeat. In such cases a trinucleotide run is expanded and leads to pathogenic polyglutamine tracts. In Huntington disease, expansion of a CAG repeat (polyglutamine) beyond a certain threshold is pathogenic. The presence of this expanded polyglutamine produces a "toxic gain of function" in huntingtin. The concomitant protein aggregation is likely to be the cause of neuronal death associated with inclusions of mutant protein in susceptible neurons.[72] It is now accepted that a gene may be altered in a different way from one tissue to another in the same individual or from one generation to the next. Accordingly, instability of the coding microsatellite in the huntingtin gene provides clues about the phenomenon of anticipation. Anticipation is manifested as the increasing severity, declining age of onset, and increasing penetrance of an inherited disorder as generations pass.[73] Indeed, for Huntington disease, trinucleotide amplification correlates with an earlier onset. Recently, polyalanine expansions have enlarged this group. Specifically, this has been shown for a mutation in a gene encoding an abundant nuclear protein that binds with high affinity to nascent polyA tails (polyA-binding protein 2 or PAPB2). Indeed, the autosomal dominant form of oculopharyngeal muscular dystrophy (OPMD), an adult-onset disease, is caused by short expansions of the polyAlanine tract of PABP2. Calado et al (2000)[74] have reported that PABP2 appears in filamentous nuclear inclusions, which are the pathological hallmark of OPMD. These inclusions also contain ubiquitin and subunits of the proteasome. Insoluble PABP2 is more resistant to salt extraction than the protein dispersed in the nucleoplasm. This suggests that the polyalanine expansions in PABP2 induce misfolding and aggregation of the protein, similar to events in neurodegenerative diseases caused by polyglutamine expansions.

New Protein Functions/Chimeric Proteins

The most prominent examples in the class are cancer-associated fusion proteins. They have been especially well studied in leukemias and may impart an abnormal phenotype through common modes of transcriptional dysregulation. The chimeric proteins are generated by genomic translocations and usually contain a DNA-binding domain, domains responsible for homo- or hetero-dimerization and domains that interact with proteins involved in chromatin remodeling (e.g., corepressor or coactivator molecules). These shared features constitute "variations on the theme" that underlies the aberrant growth and differentiation that is the hallmark of leukemia cells.[75]

Apparent Dominant Phenotype with Recessivity at the Cellular Level

The best known paradigm of this class concerns the retinoblastoma gene RB1, a tumor suppressor gene (TSG). At the cellular level the presence of only one copy of RB1 is compatible with a normal proliferation phenotype. However, in a population of cells, somatic mutations can obliterate the active copy in a cell leading to the generation of a malignant clone that may form a tumor. This recapitulates the classical "second hit" theory promoted by Knudson.[76,77] In this mechanism, a germ line mutation of one allele predisposes cells to tumor formation, while an additional somatic mutation of the other allele causes malignant transformation. For this to occur, the double mutant must have a growth advantage over the wild-type. Familial cases of Wilms' tumor (mentioned above) with mutations in WT-1 fit also Knudson's model.[62] Note however that there are cases, such as that of AML1/RUNX1, where malignant cells do not show any loss of the remaining active AML1 allele.[78] This suggests that HI of a TSG itself may also play a role in tumor formation. It is also likely that monoallelic expression and particularly imprinting may play a role at least in some cases.

Nonalellic Noncomplementation

This situation is also termed unlinked noncomplementation and arises when recessive mutations in two different loci fail to complement each other, that is, the double heterozygote exhibits a phenotype. Thus, mutations in one locus appear to be dominant given the individuals are heterozygous for mutations in the other, predisposing, locus. This phenomenon has been seen: (i) in physically interacting products, (ii) in proteins that do not interact directly but belong to the same macromolecular complex, (iii) between gene products acting in the same pathway but also (iv) in proteins that are not in the same pathway.[79] Two models have been proposed to explain non alellic non complementation: the dosage and the poison models.[80,81] In the former only a decrease in the dosage of both loci will lead to an abnormal phenotype. This is the result of a combined HI of the two loci. In the poison model, a mutant product may bind to its partner without producing a phenotype. A second mutation in the partner will in turn lead to an abnormal phenotype. For example some mutations in the alpha and beta tubulins in *Drosophila* or yeast fail to complement each other.[82] Many other examples have been reported in transcriptional regulation, cytoskeletal components, neuron outgrowth and oogenesis.[79]

In its strongest form, this type of epistasis can lead to the lethality of particular allelic combinations that will only be revealed via crossing and the recombination of genotypes that separately harbor one of the deleterious alleles. This leads to the synthetic lethals, or more generally, synthetic deleterious loci. The biological relevance of synthetic lethals has been controversial, but studies show that the equilibrium frequencies of these deleterious alleles can be relatively high because both dominance and epistasis act to shield carriers from exposure to selection. The frequencies of synthetic deleterious alleles depend on the mutation rate at both loci, on how recessive the mutant allele is relative to the wild-type allele in heterozygotes, on the strength of selection against the synthetic genotype as well as the genotypes carrying the mutations and on the rate of recombination between the interacting loci.[83]

Polar Overdominance

The callipyge phenotype described in sheep is an inherited muscular hypertrophy that is subject to an unusual parent-of-origin (nonMendelian) effect referred to as polar overdominance. Only heterozygous individuals having inherited the mutation from their father exhibit the muscular hypertrophy.[84] This is the only known case in mammals of paternal polar overdominance. The causal mutation is a single base pair change[85] that enhances expression of coregulated imprinted genes in *cis* without affecting their imprinting status.[86] Generation of the phenotype may also involve *trans* interactions between the products of reciprocally imprinted genes.[87] The mutation lies in a region of high homology among mouse, sheep, cattle, and humans, but not in any previously identified expressed transcript and this region contains a long probably a long-range control element within this imprinted domain.

Dosage Effects and the Dosage Balance Notion

When we talk about dominance linked to HI or increased gene expression, we think of changes of absolute expression levels. However, it has been recently suggested that dosage balance is a key issue.[46] Let us take the example of Mlc1p which is a light chain for the myosin Myo2p in the budding yeast. Mlc1p stabilizes Myo2p and exhibits HI. However, reduced amounts of Myo2p help overcome HI.[88] A relative excess of Myo2p is thought to be responsible for a "toxic effect". It is increasingly clear that the stoichiometric balance and not only the absolute levels of proteins plays a more important role than previously supposed in the etiology of dominance. Consider the case of HI of PLP which leads to a mild disease that results in widespread demyelination in the nervous system while HI of PMP22 leads to a hereditary neuropathy with liability to pressure palsies. However, the over-expression of either protein also leads to abnormal dominant phenotypes, as mentioned in a previous section.[53] In man, diseases caused by increased or decreased gene dosage due to genomic rearrangments have been referred to as "genomic disorders" by Lupski.[89] It is obvious that, in addition to genomic causes, dosage effects may arise from many other types of mutations including those that alter the half lives of the mRNAs or the proteins they encode, as well as regulatory mutations.

Let us analyze dosage balance within a macromolecular complex on more theoretical grounds. The components of a macromolecular complex (proteins or RNAs) are not all identical with respect to their location, interactions in the context of a complex and potential dosage sensitivity (more details in Chapter 8).[90,91] For instance, a subunit may form a single bridge between two or more separable parts of a complex (such as B in A-B-C). A modest increase in the concentration of this bridging subunit may induce a drastic inhibition of complex formation. An excess of the B component of the A-B-C complex may lead to "trapped" inactive subcomplexes AB and BC in irreversible conditions. Considerations of similar nature lead to the dosage balance theory which proposes that stoichiometric imbalances in macromolecular complexes can be a source of dominant phenotypes.

The dosage balance theory predicts that:

- adaptations should tend to minimize the degree of imbalance,
- heterozygous deletions or over-expression of one subunit should be deleterious,
- the strength of transcriptional coregulation of subunits is expected to reflect dosage sensitivity
- and an imbalance caused by halving (or increasing) gene dosage in one gene will be rescued, at least in part, by reducing (or increasing) expression of the interacting partner.

The DBT has received strong statistical support in a recent analysis of yeast genomic data. According to the MIPS Comprehensive Yeast Genome Database catalogue, approximately 30% of yeast genes code for proteins involved in annotated complexes. Focusing on essential genes (i.e., those where a homozygote deletion is lethal) Papp et al[92] have shown that dosage sensitive genes (low heterozygote fitness) are at least two times more likely to encode proteins involved in complexes than genes with low dosage sensitivity. They also observe that when considering genes that encode proteins that are only involved in one complex, heterozygote fitness shows a weak negative correlation with the size of the complex. However, the number of protein complexes in which a gene is involved has no detectable "effect" on heterozygote fitness. Being in complex appears to be the most important factor affecting dosage sensitivity. Furthermore, about 50% of the genes whose over-expression in wild-type cells is lethal are members of a protein complex. This is a highly-significant excess as compared to genes inducing no clear detrimental effect on fitness when over-expressed. For the subunit pairs with relatively high fitness deficiency (>15 %), more than 80% of them are tightly coexpressed. These authors argue that the above evidence supports the notion that dominance might be a by-product of physiological mechanisms, rather than an adaptation to mask the harmful effects of mutations. However, one cannot neglect the possibility of selectable strategies to attain and preserve the right stoichiometry in a complex or palliate its alterations, i.e., modification of dominance.[91]

Another prediction of the DBT relevant in the context of dominance is that single gene duplications of certain types of subunits can be harmful, because it may cause imbalance. Thus, genes encoding interacting pairs should remain as single copies or both should undergo gene duplication. When Papp et al[92] analyzed pairs of genes coding for interacting subunits, they find a large excess of pairs with the same number of paralogs and that genes belonging to huge families seldom encode complex components.

Problems of dosage imbalance are not restricted to yeast, but higher organisms may have other sources of dominance arising from complex developmental and signaling pathways. As pointed out by Birchler et al,[93] stoichiometric imbalances in gene regulatory cascades may explain the dominant effects of aneuploidy. A formal extension of the DBT to biochemical, signaling and genetic pathways is straightforward (Veitia, *Genetics*, in press).

Dominance and Robustness against Deleterious Effects from a Global Perspective

Total deletion of a gene may have a small phenotypic effect, owing to compensation. This is due either to the existence of duplicate genes or of alternative metabolic/signaling pathways. Gu et al[94] have shown that, in yeast, there is a significantly higher probability of compensation for a gene with two recognizable copies than for a singleton. They found a high correlation between the frequency of compensation and the sequence similarity of the two duplicates. Furthermore, a higher probability of a severe fitness effect was also reported when the more highly expressed copy was deleted. The authors estimate that duplicates compensate for at least a quarter of those gene deletions which have no phenotype.

The ability of a duplicate to compensate for the deletion of the other duplicate is critically dependent on their subfunctionalization; that is whether each copy of the duplicate gene has acquired or retains complementary subfunctions.[95] Subfunctionalization may simply lead to a reduction of the levels of expression of both the original and the copy. Thus, a diploid organism would have four copies coding for essentially the same protein, which clearly imparts robustness in cases of heterozygous deletions or duplications of only one copy. One would predict, therefore, that an increase in the number of subfunctionalized copies will increase robustness, not only against deletion, but also against partial duplications. The work by Gu et al[94] supports the notion that a massive subfunctionalization in yeast following very-likely a genome-wide duplication is responsible for a resilience against heterozygous and apparently null mutations.

Another interesting agent leading to robustness against mutations is the heat-shock protein 90 (Hsp90) which chaperones the maturation of many regulatory proteins. Indeed, Hsp90 mutations has been shown to uncover genetic variation that exists in a silent state in *Drosophila melanogaster*. Thus, Hsp90 is thought to buffer genetic variation in morphogenetic pathways, allowing them to accumulate variants under neutral conditions.[96] More recently, similar results have been obtained in *Arabidopsis*. Impairment of Hsp90 function produces a panoply of morphological phenotypes, which are dependent on underlying genetic variation. In short, Hsp90 influences morphogenetic responses to environmental cues and buffers normal development from the destabilizing effects of stochastic processes.[97] This is a typical example of canalization: the phenotypic insensitivity to mutation in developmental pathways.

On the other hand, many recent publications suggest that at the topological level, metabolic and protein-protein interaction networks are organized in such a way that a few components interact with many other (highly connected nodes or hubs).[98,99] This highly inhomogeneous connectivity is thought to impart resilience against deleterious mutations to the network, as most of the nodes are poorly connected. It has been proposed that this inhomogeneity has been selected for.[100] Alternatively, it is conceivable that this topology is a main evolutionary result of the addition of new nodes (end-users, in internet terms) onto the ancient ones, that will tend to become the hubs (see ref. 101). But of course selection is expected to have played a role in shaping the present architecture of the networks. For reasons of simplicity, the pure topological approach to gene networks and its conclusions do not take into account the dynamics (changes in concentration) of the components. Furthermore, only "hard links" (strong/stable interactions) between the nodes have been considered so far. Most authors claim that the inhomogeneous nature of the network makes it resistant to random errors, because most mutations will affect poorly connected nodes (by far more abundant). However, this does not imply at all that there will be no phenotype. What is likely is that the phenotype will be less pleiotropic than that following removal of a hub. In this context, it is also worth taking into account that all the links between the nodes may not have the same strength. The links that are more easily detectable with our current experimental methods are the strongest ones (hard links). They are mainly responsible for the inhomogeneous character of networks. In probabilistic terms, hard links would have a probability of the interaction between the nodes close to one while weaker links will have lower probabilities and would, thus, be more difficult to uncover. These "probabilistic weights" are likely to change according to the particular configuration of the network (i.e., when all nodes are present versus when some are lacking). The introduction of weighted networks, is likely to revolutionize our "hard-wired" topological perception of gene interactions. Developmental robustness (canalization) is thought to evolve due to long-term natural selection for optimal phenotypes. However, developmental processes, modeled as a network of interacting transcriptional regulators, generate canalization even without selection toward an optimum. The extent of canalization, measured as the insensitivity of a network in the equilibrium state to mutations, depends on the complexity of the network. More highly connected networks evolve to be more canalized. Thus, canalization would be an inevitable consequence of complex developmental-genetic processes and would not require explanation in terms of evolution to suppress phenotypic variation such as the buffer effect of Hsp90.[102] This can be matter of debate.

Concluding Remarks

Mendel made a precise description of dominance in 1866. Seventy years later, Wright postulated the first mechanistic explanation based on biochemical analyses. Wright's model contradicted Fisher's theory of "modifiers", leading to a long-standing controversy. It took fifty years more before Kacser and Burns proposed a simple explanation of dominance that is restricted to metabolic systems. However, it was only with the advent of molecular biology that we started to get mechanistic insights. The study of metabolic, developmental or transcriptional systems has reinforced the notion that nonlinearity in the relationship between genotypic and phenotypic values, which are structural characteristics of biological systems, are a source of dominance. However, the mechanisms generating nonlinear behavior are diverse. It is likely that dominance is a systemic property of biological systems, as suggested by Fisher's oponents. However, we can not neglect the study of evolutionary pressures that influence dominance relationships. Although on more solid grounds than in the past century the debate about the sources, mechanisms and evolution of dominance is still going on.

Acknowlegdements

The authors thank D. de Vienne D, C. Dillmann and B. Lakowski for helpful comments on the manuscript.

References

1. Mendel G. Versuche über pflanzenhybriden. Verhandlung des naturforschenden Vereines in Brünn 1866; 4:3–47.
2. Kacser H, Burns JA. The molecular basis of dominance. Genetics 1981; 97(3-4):639-666.
3. Miller WJ. Dominance. Braz J Genet 1996; 19(4):677.
4. Miller WJ. Dominance, codominance and epistasis. Braz J Genet 1997; 20(4):663-665.
5. Grüneberg H. The genetics of the mouse. The Hague: Martinus Nijhoff, 1992.
6. Wright S. Posnatal changes in the intensity of coat color in diverse genotypes of the guinea pig. Genetics 1960; 45:1503-1529.
7. Müller HJ. Further studies on the nature and causes of gene mutations. In: Jones DF, ed. Proceedings of the Sixth International Congress on Genetics. Menasha: Banta, 1932:2:213-255.
8. Fisher RA. The possible modification of the response of the wild type to recurrent mutations. Am Nat 1928; 62:115-126.
9. Jaenisch R. Transgenic animals. Science 1988; 240(4858):1468-1474.
10. Mackay TFC, Lyman R, Jackson MS. Effects of P elements on quantitative traits in Drosophila melanogaster. Genetics 1992; 130(2):315-332.
11. Mukai T, Chigusa SI, Mettler LE et al. Mutation rate and dominance of genes affecting viability in Drosophila melanogaster. Genetics 1972; 72(2):333-355.
12. Gilchrist MA, Nijhout HF. Nonlinear developmental processes as sources of dominance. Genetics 2001; 159(1):423-432.
13. Wright S. Molecular and evolutionary theories of dominance. Am Nat 1934; 68:24-53.
14. Falconer DS, Mackay TFC. Introduction to quantitative genetics. 4th ed. London: Longman, 1996:356-378.
15. Bost B, Dillmann C, de Vienne D. Fluxes and metabolic pools as model traits for quantitative genetics. I. The L-shaped distribution of gene effects. Genetics 1999; 153(4):2001-2012.
16. Cheverud JM, Routman EJ. Epistasis and its contribution to genetic variance components. Genetics 1995; 139(3):1455-1461.
17. Dillmann C, Foulley JL. Another look at multiplicative models in quantitative genetics. Genet Select Evol 1998 30:543-564.
18. Omholt SW, Plahte E, Oyehaug L et al. Gene regulatory networks generating the phenomena of additivity, dominance and epistasis. Genetics 2000; 155(2):969-980.
19. Fisher RA. The evolution of dominance. Biol Rev Camb Philos Soc 1931; 6:345-368.
20. Fisher RA. The genetical theory of natural selection. 2nd ed. New York: Dover Publications, 1958.
21. Crosby JL. The evolution and nature of dominance. J Theor Biol 1963; 5(1):35-51.
22. Charlesworth B. Evidence against Fisher's theory of dominance. Nature (London) 1979; 278:848-849.
23. Orr HA. A test of Fisher's theory of dominance. Proc Natl Acad Sci USA 1991; 88(24):11413-11415.
24. Kacser H, Burns JA. The control of flux. Symp Soc Exp Biol 1973; 27:65-104.
25. Heinrich R, Rapoport TA. Linear theory of enzymatic chains; its application for the analysis of the crossover theorem and of the glycolysis of human erythrocytes. Acta Biol Med Ger 1973; 31(4):479-494.
26. Cornish-Bowden A. Dominance is not inevitable. J Theor Biol 1987; 125(3):333-338.
27. Grossniklaus U, Madhusudhan MS, Nanjundiah V. Nonlinear enzyme kinetics can lead to high metabolic flux control coefficients: Implications for the evolution of dominance. J Theor Biol 1996; 182(3):299-302.
28. Bagheri-Chaichian H, Hermisson J, Vaisnys JR et al. Effects of epistasis on phenotypic robustness in metabolic pathways. Math Biosci 2003; 184(1):27-51.
29. Porteous JW. Dominance—one hundred and fifteen years after Mendel's paper. J Theor Biol 1996; 182(3):223-232.
30. Jimenez-Sanchez G, Childs B, Valle D. Human disease genes. Nature 2001; 409(6822):853-855.
31. Nanjundiah V. Why are most mutations recessive. J Genet 1993; 72(2/3):85-97.
32. van Dooren TJM. The evolutionary ecology of dominance-recessivity. J Theor Biol 1999; 198(4):519-532.
33. Bürger R. Dynamics of the classical genetic model for the evolution of dominance. Math Biosci 1983; 67(2):125-143.

34. Wagner GP, Bürger R. On the evolution of dominance modifiers. II. A nonequilibrium approach to the evolution of genetic systems. J Theor Biol 1985; 113(3):475-500.
35. Mayo O, Bürger R. The evolution of dominance: A theory whose time has passed? Biol Rev Camb Philos Soc 1997; 72:97-110.
36. Todd JA. From genome to aetiology in a multifactorial disease, type 1 diabetes. Bioessays 1999; 21(2):164-174.
37. Katsanis N, Ansley SJ, Badano JL et al. Triallelic inheritance in Bardet-Biedl syndrome, a Mendelian recessive disorder. Science 2001; 293(5538):2256-2259.
38. Wilkie AOM. The molecular basis of genetic dominance. J Med Genet 1994; 31(2):89-98.
39. Lamande SR, Bateman JF, Hutchison W et al. Reduced collagen VI causes Bethlem myopathy: A heterozygous COL6A1 nonsense mutation results in mRNA decay and functional haploinsufficiency. Hum Mol Genet 1998; 7(6):981-989.
40. Richards AJ, Baguley DM, Yates JRW et al. Variation in the vitreous phenotype of Stickler syndrome can be caused by different amino acid substitutions in the X position of the type II collagen gly-X-Y triple helix. Am J Hum Genet 2000; 67(5):1083-1094.
41. Na GC, Butz LJ, Bailey DG et al. In vitro collagen fibril assembly in glycerol solution: Evidence for a helical cooperative mechanism involving microfibrils. Biochemistry 1986; 25(5):958-966.
42. Tassabehji M, Metcalfe K, Hurst J et al. An elastin gene mutation producing abnormal tropoelastin and abnormal elastic fibres in a patient with autosomal dominant cutis laxa. Hum Mol Genet 1998; 7(6):1021-1028.
43. Toonkool P, Regan DG, Kuchel PW et al. Thermodynamic and hydrodynamic properties of human tropoelastin. analytical ultracentrifuge and pulsed field-gradient spin-echo nmr studies. J Biol Chem 2001; 276(30):28042-28050.
44. Cho C, Willis WD, Goulding EH et al. Haploinsufficiency of protamine-1 or -2 causes infertility in mice. Nat Genet 2001; 28(1):82-86.
45. Willmitzer L, Wagner KG. The binding of protamines to DNA; role of protamine phosphorylation. Biophys Struct Mech 1980; 6(2):95-110.
46. Veitia RA. Exploring the etiology of haploinsufficiency. Bioessays 2002; 24(2):175-184.
47. Seidman JG, Seidman C. Transcription factor haploinsufficiency: When half a loaf is not enough. J Clin Invest 2002; 109(4):451-415.
48. Veitia RA. A sigmoidal transcriptional response: Cooperativity, synergy and dosage effects. Biol Rev Camb Philos Soc 2003; 78(1):149-170.
49. Cook DL, Gerber AN, Tapscott SJ. Modeling stochastic gene expression: Implications for haploinsufficiency. Proc Natl Acad Sci USA 1998; 95(26):15641-15646.
50. Weinstein LS. The role of tissue-specific imprinting as a source of phenotypic heterogeneity in human disease. Biol Psychiatry 2001; 50(12):927-931.
51. Nutt SL, Busslinger M. Monoallelic expression of Pax5: A paradigm for the haploinsufficiency of mammalian Pax genes? Biol Chem 1999; 380(6):601-611.
52. Tschudy DP, Valsamis M, Magnussen CR. Acute intermittent porphyria: Clinical and selected research aspects. Ann Intern Med 1975; 83(6):851-864.
53. Woodward K, Malcolm S. CNS myelination and PLP gene dosage. Pharmacogenomics 2001; 2(3):263-272.
54. Foster JW, Dominguez-Steglich MA, Guioli S et al. Campomelic Dysplasia and autosomal sex reversal caused by mutations in an SRY-related gene. Nature 1994; 372(6506):525-529.
55. Huang B, Wang S, Ning Y et al. Autosomal XX sex reversal caused by duplication of SOX9. Am J Med Genet 1999; 87(4):349-353.
56. Bishop CE, Whitworth DJ, Qin Y et al. A transgenic insertion upstream of sox9 is associated with dominant XX sex reversal in the mouse. Nat Genet 2000; 26(4):490-494.
57. Sun XH. Constitutive expression of the Id1 gene impairs mouse B cell development. Cell 1994; 79(5):893-900.
58. Becker MA, Kostel PJ, Meyer LJ et al. Human phosphoribosylpyrophosphate synthetase: Increased enzyme specific activity in a family with gout and excessive purine synthesis. Proc Natl Acad Sci USA 1973; 70(10):2749-2752.
59. Beitel GJ, Clark SG, Horvitz HR. Caenorhabditis elegans ras gene let-60 acts as a switch in the pathway of vulval induction. Nature 1990; 348(6301):503-509.

60. Weinstein LS, Shenker A, Gejman PV et al. Activating mutations of the stimulatory G protein in the McCune-Albright syndrome. N Engl J Med 1991; 325(24):1688-1695.
61. Herskowitz I. Functional inactivation of genes by dominant negative mutations. Nature 1987; 329(6136):219-222.
62. Re GG, Hazen-Martin DJ, Sens DA et al. Nephroblastoma (Wilms' tumor): A model system of aberrant renal development. Semin Diagn Pathol 1994; 11(2):126-135.
63. Parant JM, Lozano G. Disrupting TP53 in mouse models of human cancers. Hum Mutat 2003; 21(3):321-326.
64. Webster LC, Ricciardi RP. trans-dominant mutants of E1A provide genetic evidence that the zinc finger of the trans-activating domain binds a transcription factor. Mol Cell Biol 1991; 11(9):4287-4296.
65. Heger P, Rosorius O, Hauber J et al. Titration of cellular export factors, but not heteromultimerization, is the molecular mechanism of trans-dominant HTLV-1 rex mutants. Oncogene 1999; 18(28):4080-4090.
66. Robinson PN, Godfrey M. The molecular genetics of Marfan syndrome and related microfibrillopathies. J Med Genet 2000; 37(1):9-25.
67. Smith F. The molecular genetics of keratin disorders.Am J Clin Dermatol 2003; 4(5):347-364.
68. Martin DI, Tsai SF, Orkin SH. Increased gamma-globin expression in a nondeletion HPFH mediated by an erythroid-specific DNA-binding factor. Nature 1989; 338(6214):435-438.
69. Forget BG. Molecular basis of hereditary persistence of fetal hemoglobin. Ann N Y Acad Sci 1998; 850:38-44.
70. Whiting J. Craniofacial abnormalities induced by the ectopic expression of homeobox genes. Mutat Res 1997; 396(1-2):97-112.
71. Buxbaum JN, Tagoe CE. The genetics of the amyloidoses. Annu Rev Med 2000; 51:543-569.
72. Hoffner G, Djian P. Protein aggregation in Huntington's disease. Biochimie 2002; 84(4):273-278
73. Carpenter NJ. Genetic anticipation. Expanding tandem repeats. Neurol Clin 1994; 12:683-697.
74. Calado A, Tome FM, Brais B et al. Nuclear inclusions in oculopharyngeal muscular dystrophy consist of poly(A) binding protein 2 aggregates which sequester poly(A) RNA. Hum Mol Genet 2000; 9(15):2321-2328.
75. Scandura JM, Boccuni P, Cammenga J et al. Transcription factor fusions in acute leukemia: Variations on a theme. Oncogene 2002; 21(21):3422-3444.
76. Knudson Jr AG. Mutation and cancer: Statistical study of retinoblastoma. Proc Natl Acad Sci USA. 1971; 68(4):820-823.
77. Knudson AG. Hereditary cancer: Two hits revisited. J Cancer Res Clin Oncol 1996; 122(3):135-140.
78. Largaespada DA. Haploinsufficiency for tumor suppression: The hazards of being single and living a long time. J Exp Med 2001; 193(4):F15-18.
79. Yook KJ, Proulx SR, Jorgensen EM. Rules of nonallelic noncomplementation at the synapse in Caenorhabditis elegans. Genetics 2001; 158(1):209-220.
80. Fuller MT, Regan CL, Green LL et al. Interacting genes identify interacting proteins involved in microtubule function in Drosophila. Cell Motil Cytoskeleton 1989; 14(1):128-135.
81. Stearns T, Botstein D. Unlinked noncomplementation: Isolation of new conditional-lethal mutations in each of the tubulin genes of Saccharomyces cerevisiae. Genetics 1988; 119(2):249-260.
82. Hays TS, Deuring R, Robertson B et al. Interacting proteins identified by genetic interactions: A missense mutation in alpha-tubulin fails to complement alleles of the testis-specific beta-tubulin gene of Drosophila melanogaster. Mol Cell Biol 1989; 9(3):875-884.
83. Phillips PC, Johnson NA. The population genetics of synthetic lethals. Genetics 1998; 150(1):449-458.
84. Cockett NE, Jackson SP, Shay TL et al. Polar overdominance at the ovine callipyge locus. Science 1996; 273(5272):236-238.
85. Freking BA, Murphy SK, Wylie AA et al. Identification of the single base change causing the callipyge muscle hypertrophy phenotype, the only known example of polar overdominance in mammals. Genome Res 2002; 12(10):1496-1506.
86. Charlier C, Segers K, Karim L et al. The callipyge mutation enhances the expression of coregulated imprinted genes in cis without affecting their imprinting status. Nat Genet 2001; 27(4):367-369.

87. Georges M, Charlier C, Cockett N. The callipyge locus: Evidence for the trans interaction of reciprocally imprinted genes. Trends Genet 2003; 19(5):248-252.
88. Stevens RC, Davis TN. Mlc1p is a light chain for the unconventional myosin Myo2p in Saccharomyces cerevisiae. J Cell Biol 1998; 142(3):711-722.
89. Lupski JR. Genomic disorders: Structural features of the genome can lead to DNA rearrangements and human disease traits. Trends Genet 1998; 14(10):417-422.
90. Bray D, Lay S. Computer-based analysis of the binding steps in protein complex formation. Proc Natl Acad Sci USA 1997; 94(25):13493-13498.
91. Veitia RA. Nonlinear effects in macromolecular assembly and dosage sensitivity. J Theor Biol 2003; 220(1):19-25.
92. Papp B, Pal C, Hurst LD. Dosage sensitivity and the evolution of gene families in yeast. Nature 2003; 424(6945):194-197.
93. Birchler JA, Bhadra U, Bhadra MP et al. Dosage-dependent gene regulation in multicellular eukaryotes: Implications for dosage compensation, aneuploid syndromes, and quantitative traits. Dev Biol 2001; 234(2):275-288.
94. Gu Z, Steinmetz LM, Gu X et al. Role of duplicate genes in genetic robustness against null mutations. Nature 2003; 421(6918):63-66.
95. Lynch M, O'Hely M, Walsh B et al. The probability of preservation of a newly arisen gene duplicate. Genetics 2001; 159(4):1789-1804.
96. Rutherford SL, Lindquist S. Hsp90 as a capacitor for morphological evolution. Nature 1998; 396(6709):336-342.
97. Queitsch C, Sangster TA, Lindquist S. Hsp90 as a capacitor of phenotypic variation. Nature 2002; 417(6889):618-624.
98. Jeong H, Tombor B, Albert R et al. The large-scale organization of metabolic networks. Nature 2000; 407(6804):651-654.
99. Jeong H, Mason SP, Barabasi AL et al. Lethality and centrality in protein networks. Nature 2001; 411(6833):41-42.
100. Albert R, Jeong H, Barabasi AL. Error and attack tolerance of complex networks. Nature 2000; 406(6794):378-382.
101. Wilkins AS. Moving up the hierarchy: A hypothesis on the evolution of a genetic sex determination pathway. Bioessays 1995; 17(1):71-77.
102. Siegal ML, Bergman A. Waddington's canalization revisited: Developmental stability and evolution. Proc Natl Acad Sci USA 2002; 99(16):10528-10532.
103. Goldstein JL, Brown MS. Atherosclerosis: The low-density lipoprotein receptor hypothesis. Metabolism 1977; 26(11):1257-75.
104. Vergopoulos A, Bajari T, Jouma M et al. A xanthomatosis-susceptibility gene may exist in a Syrian family with familial hypercholesterolemia. Europ J Hum Genet 1997; 5(5):315-323.
105. Ekstrom U, Abrahamson M, Floren CH et al. An individual with a healthy phenotype in spite of a pathogenic LDL receptor mutation (C240F). Clin Genet 1999; 55(5):332-339.

CHAPTER 3

Dominance, Nonlinear Developmental Mapping and Developmental Stability

Christian Peter Klingenberg

Abstract

Developmental stability is the ability of organisms to buffer against the random variation that arises spontaneously as a consequence of stochastic variation in the cellular processes that are involved in the development of morphological structures. Its converse, developmental instability, is the imprecision that leads to morphological variability even when genetic and environmental conditions are kept constant, and can be conveniently measured as the random left-right differences of bilaterally symmetric organisms. This chapter demonstrates that the genetic control of developmental stability is intimately connected with nonadditive genetic variation of the morphological traits of interest. Dominance and epistasis have also been shown by empirical studies to play an important role in the genetic architecture of developmental stability. A brief review of some mechanisms that generate stochastic variation in gene expression suggests that nonadditive genetic variation is also an important factor for the origin of developmental noise.

Introduction

The molecular and cellular processes that constitute development are inherently variable, yet they contribute to the reliable assembly of intricately organized body plans. The mechanisms that achieve this reliability at the level of the phenotype are referred to as developmental stability. The nature of these mechanisms is not well understood, and research going beyond purely phenomenological studies has only started relatively recently.[1] Developmental stability is subset of a broader class of developmental buffering phenomena, which also includes canalization against genetic and environmental effects.[2,3] It has been contentious whether different mechanisms are involved in developmental stability and canalization[4,5] or whether both are due to the effects of the same system on different kinds of variation.[6,7]

It is also an open question to what extent developmental buffering results from the action of special mechanisms or from robustness that is a consequence of the setup of developmental mechanisms. The most prominent case in favor of special mechanisms is the case of the Hsp90 chaperone protein, where the protein has a clear function in the maintenance of cellular function and inhibition of its action results in increased phenotypic variation and higher incidences of overt anomalies of development.[8-10] It is unclear, however, how far this example can be generalized, as a variety of developmental systems have been shown to be inherently robust against intrinsic and extrinsic perturbations by virtue of the way their components interact.[11,12] It is not evident whether buffering is a result of special mechanisms that have evolved as an adaptation for this function[13] or whether it arose as an automatic attribute of developmental

The Biology of Genetic Dominance, edited by Reiner A. Veitia.

systems, although buffering is clearly evolvable to a certain degree. This question is a parallel to the debate on whether dominance is an evolved property of genes or an automatic result of the biochemical and physiological function of genetic systems.[14-18]

In this chapter, I will primarily explore those origins of developmental buffering that result from the general setup of developmental systems, because they are directly linked to the issue of dominance and haploinsufficiency. Because this argument relies primarily on the nonlinear relations between inputs and outputs of developmental systems, which is a very widespread property of biochemical and developmental processes, it is likely that the conclusions apply to a broad range of organisms and developmental contexts. I also briefly review the literature on the molecular origins and implications of developmental noise, which may itself relate to dominance and haploinsufficiency.

Developmental Stability and Its Measurement

Developmental stability is the ability of developmental processes to resist fluctuations of the system and produce a consistent phenotype according to the genotype and environment of the organism.[1] This degree of resistance against possible perturbations is inherently difficult to measure, but it is easier to quantify its opposite, developmental instability, which is the imprecision of a developmental system and its morphological end products. It can be quantified by the amount of variation among phenotypes that would be produced by the same developmental process run repeatedly under identical genetic and environmental conditions, or in a situation close to this ideal.

In bilaterally symmetric organisms or parts, fluctuating asymmetry,[1,19,20] the random differences between left and right sides, offers an easy means to study developmental instability. Both sides share the same genome (barring somatic mutations) and usually develop under nearly identical environmental conditions, and therefore the variation of asymmetry around its average is due to random fluctuations of developmental processes, and can be used as a measure of developmental instability.[1] This measure of developmental instability is a composite of the opposing effects of developmental noise and the system's capacity to buffer against it.

Similar reasoning has been applied at the cellular scale to quantify the relative contributions to stochastic variation of gene expression that are intrinsic and extrinsic to individual genes.[21,22] These components of variation can be separated by studying the expression of two identical genes in the same cell, for instance, two constructs with different varieties of a reporter gene such as green fluorescent protein.[22] Random variation that is intrinsic to a gene can be isolated from the contrast between the expression levels of the two genes, whereas the extrinsic component of noise can be estimated from the variation that is correlated between the two copies.[21,22] The logic of the comparison of the behavior of two genes in a cell corresponds to the comparison between the left and right sides of an organism in studies of fluctuating asymmetry.[1]

Developmental Mapping and Nonadditive Genetic Effects

Development has often been characterized by the relationship between the genotype and phenotype, described mathematically as some type of mapping function.[23,24] This approach can be extended to a consideration of both genetic and environmental contributions to the phenotype through a more general form of developmental mapping in which the phenotype is expressed as a function of some developmental parameter that is controlled jointly by genetic and other factors. The basis of this approach is a developmental or biochemical model, such as a diffusion-threshold process,[12,25,26] models of flux in a pathway of several enzymes,[14] or of the formation of molecular complexes.[27] For most developmental systems, these relationships are markedly nonlinear, and this nonlinearity is of critical importance for the dynamical behavior of these systems.

At the core of this approach is a dissection of the relationship between genetic and other inputs into the developmental system and the morphological output into two distinct layers or phenotypes: the first level is a developmental phenotype, which can be conceptualized as the parameters of a developmental model, and the second level is the morphological phenotype that results from the model with those parameter settings. The benefit of introducing an intermediate developmental phenotype between the genetic factors and the morphological outcome is that the genetic control of the developmental phenotype has a relatively simple basis, even in complex developmental models. For instance, if the developmental parameter corresponds to the dose of a protein, it is possible to assume an additive genetic basis. The second layer of the model considers these parameters jointly with the nongenetic factors that influence development, and thereby takes into account the epigenetic and gene-by-environment interactions between different components of the developmental model. The genetic behavior of the morphological phenotype can then be examined by combining the information about the known genotypes and the morphological output of the system.[12,25]

In a developmental process that consists of several sequential steps, the aggregate result of all effects will be passed from one step to the next without reference to their cause. Regardless of whether a given change in the activity of, say, a molecular signal is caused by a specific allele, by an environmental influence or even by a random fluctuation of the system, it will have the same effect on the subsequent step where the signal is received. This sort of process therefore provides opportunities for interactions among the effects of different origins.

The nonlinearity of developmental mapping functions is critical for the origin of nonadditive genetic effects in such systems.[12,14,25,26] Dominance at a locus, or more precisely, dominance between two alleles of a locus, is defined as a deviation of the phenotypic value of the heterozygote for the two alleles from the average of the phenotypic values for the two homozygotes.[28,29] If each allele corresponds to a particular level of activity for a component in the developmental system, such as the transcription rate for the gene or the metabolic rate of an enzyme, the phenotypic values can be graphed as a function of this activity level (Fig. 1). In this graph, dominance corresponds to a deviation of the graph from a straight line, which corresponds to completely additive gene effects. Dominance can be quantified by the degree to which the phenotypic value of the heterozygote is above or below the value midway between those for the two homozygotes.[28] In other words, the degree of dominance for two alleles is a consequence of the extent to which the developmental mapping function is nonlinear in the range defined by those alleles.

There are different kinds of such curves, and accordingly, there are different kinds of dominance. Dominance can be positive or negative, depending on whether the phenotypic value for the heterozygote is greater or less than the value intermediate between the two homozygotes, or correspondingly, whether the developmental mapping function is convex or concave upward (Fig. 1). If the magnitude of the dominance effect exceeds the additive (linear) effect, that is, if the phenotypic value for the heterozygote is greater or less than the values for both homozygotes, there is overdominance or underdominance, corresponding to ∩-shaped or ∪-shaped mapping functions, respectively (Fig. 2). Moreover, even for monotonically increasing or decreasing functions, the shape of the mapping function can vary, which may affect the dominance patterns at the level of the phenotype. Some of these curves are steep at low values and increase toward an asymptote, such as the biochemical model of dominance based on enzyme kinetics by Kacser and Burns,[14] whereas models including cooperative binding of multiple molecules produce sigmoid curves.[27]

Depending on the location of the inflection point of a sigmoid mapping function relative to the position of the genotypes at the locus of interest, this form of curve can be the basis for haploinsufficiency (Fig. 3).[27] For mapping functions that increase in a sigmoidal fashion, haploinsuffieciency occurs if the inflection point is to the right of the developmental value of the heterozygote. In general, haploinsufficiency occurs for mapping functions where the allele

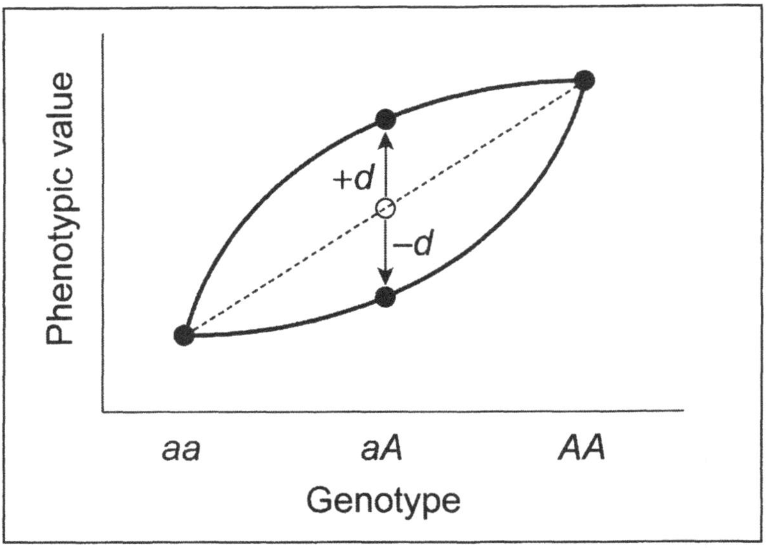

Figure 1. Nonlinear developmental mapping functions and dominance. Two developmental mapping functions are represented by the bold lines. Because they are curved, there is a difference between the phenotypic values and the midpoint of the phenotypic values for the two homozygotes. This deviation can increase or decrease the phenotypic value, corresponding to positive dominance (+*d*) or negative dominance (–*d*). The dashed line represents a linear developmental mapping function and the empty circle corresponds to the phenotypic value for completely additive effects of the gene.

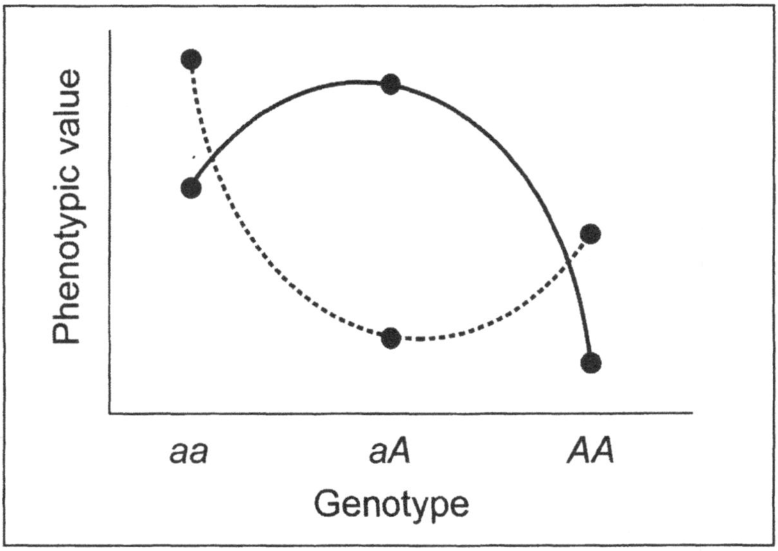

Figure 2. Overdominance and underdominance. For overdominance (solid line), the phenotypic value of the heterozygote exceeds the values of both homozygotes. For underdominance (dashed line), the phenotypic value is smaller than for either homozygote.

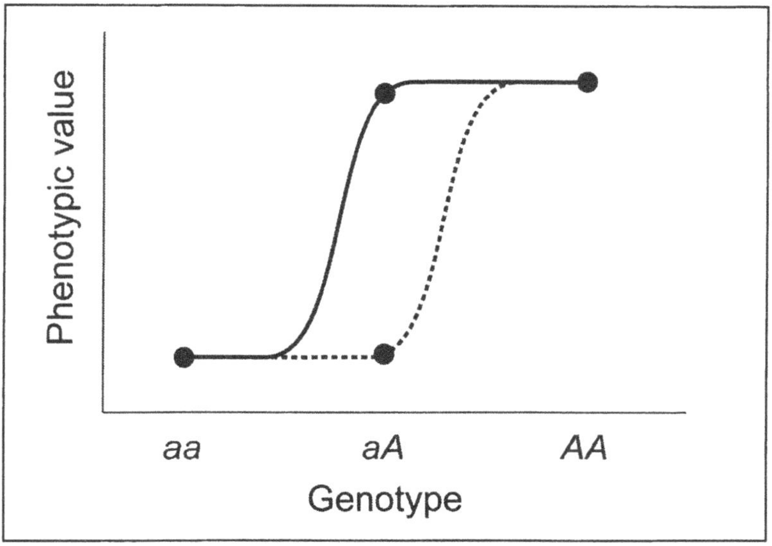

Figure 3. Sigmoidal developmental mapping functions. For these kinds of functions, the position of the genotypes along the horizontal axis is a critical determinant of the genetic behavior. The solid line shows a mapping function leading to dominance the A allele, whereas the dashed line, which has an inflection point at a higher value of the developmental variable, shows dominance of the a allele, and may be interpreted as a case of haploinsufficiency.[27]

with the lower level of activity is dominant, so that the heterozygote phenotype is similar to the "loss-of-function" phenotype. If the value for the "wild-type" phenotype is greater than the value for the "loss-of-function" allele, then any mapping function that is concave upwards will produce haploinsuffiency (e.g., Figs. 1, 3). Some caution is necessary, however, as pointed out by Veitia,[27] because it is not possible to distinguish haploinsufficiency from dominant negative effects based only on the phenotypes. Both haploinsufficiency and dominant negative effects are manifest as phenotypic changes in heterozygous condition, and therefore are dominant, but they differ in the manner in which they come about. Haploinsufficiency is the effect of reduced gene activity by mutation, where a single dose of the gene is not sufficient to produce normal developmental function and a "wild-type" phenotype.[27] In contrast, dominant negative effects are due to a mutant form of the gene product that interferes with the normal form and inhibits its function.[27] Both these phenomena are associated with similar developmental mapping functions that are concave upwards or sigmoidal (Figs. 1, 3; provided the wild-type developmental value and phenotype are scored as high values) and have similar consequences for developmental instability, and therefore need not be separated in the context of this chapter.

If more than one gene or developmental parameter is considered at the same time, the developmental mapping function becomes a surface over the space of developmental parameters. With two developmental parameters, this is a surface over the plane representing the parameters, where the elevation of the surface represents the phenotypic value. This kind of representation can easily be compared with a landscape, and various metaphors and techniques borrowed from cartography can be used to describe and visualize it. A simple way to recognize epistasis on such surfaces is to take slices through the surface parallel to the axis for one of the parameters, but set apart from each other along the axis of another parameter. The result is a comparison of developmental mapping functions for the first parameter at several distinct values for the second parameter. If these curves differ in their slope or curvature, and are not just transposed up or down, then there is an epistatic interaction between the two loci.[12,25]

Epistasis, the interactions among different loci where the genotype at one locus has an influence on the phenotypic effects of alleles at other loci,[30] is another result of the nonlinear nature of developmental mapping functions. Developmental systems are highly interactive, from the molecular and cellular level to the signals that achieve coordination of developmental processes throughout the organism, and the gene products of one locus can therefore directly affect the expression of other genes.[31,32] But even if there is no direct molecular interaction between the gene products of two loci, they still can affect each other's effects on the phenotype by changing the shapes of the respective developmental mapping functions, to produce epistasis as this term is understood in statistical genetics.[30] Such epistatic interactions among genes are almost ubiquitous, even in "textbook examples" of simple Mendelian genes,[33] and have substantial effects on genetic variation of phenotypic traits.[34-36] Nonlinear developmental models produce epistasis among the loci controlling different model parameters almost inevitably.[12,25] Normally, these interactions cannot be eliminated by any one transformation of the phenotypic variable, which distinguishes them as genuine effects from simple scaling artifacts.[28]

Nonadditive Genetic Effects and Developmental Stability

The effect of random noise in developmental processes is that the values of the developmental parameters and the phenotype are not completely determined by genetic and environmental factors, because there is variation around the expected parameter values for a particular genotype and environment. In the context of graphs like those in Figures 1–3, random noise causes deviations of the developmental parameters to the left or right from the position corresponding to a particular genotype. Accordingly, the developmental system produces a slightly modified morphological result in response to this change of the input.

The magnitude of the phenotypic response to a given small amount of random noise in the developmental system depends on the local slope of the developmental mapping function (Fig. 4). Where the mapping function is steep, a developmental perturbation will have a relatively large morphological effect, but where the mapping function is level, the same change of developmental parameters will have little morphological effect. Because developmental noise normally can be assumed to have effects that are small in comparison to the genotypic and environmental effects, a first-order approximation using the local slopes should be reasonably accurate as an estimate of the sensitivity of the system to random noise. This sensitivity to noise is a natural measure of the developmental instability of the system: its tendency to respond to developmental perturbation.[1,12]

Developmental stability, the system's ability to withstand developmental perturbations, is a consequence of the developmental functions that are relatively flat. This developmental stability does not necessarily rely on any buffering mechanisms that would actively oppose or compensate for variation in the developmental system, but the flat mapping function may indicate that some developmental parameters may not be relevant for the phenotypic response of the system under the given circumstances. This possibility has implications, for instance, for theories of developmental instability that assume the existence of metabolic or fitness costs of developmental buffering.[20] Developmental stability resulting from the structure of developmental systems may not be associated with any such costs, depending on the particular nature of the developmental system. This argument does not rule out the possibility that genuine buffering processes are involved in generating stability, for instance the well-known example of the Hsp90 chaperone protein.[8,9] Indeed, it is conceivable that the flat mapping functions themselves may be a result of buffering in some cases, but it should be emphasized that special buffering mechanisms are not the only basis for generate developmental stability.

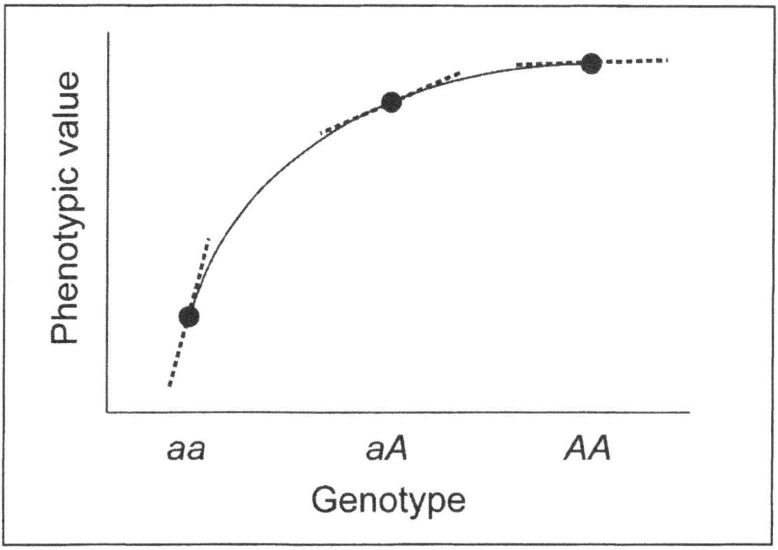

Figure 4. Slopes of the developmental mapping function at the locations corresponding to the three genotypes at a locus. These slopes are local indicators for the sensitivity of the phenotype to small perturbations of the developmental parameter, and are therefore a natural measure of developmental instability.[12] The three genotypes differ in their local slopes because the developmental mapping function is nonlinear (the A allele is partially dominant over the a allele). As a consequence, this locus has an effect on developmental instability.

For nonlinear developmental mapping functions, the local slope changes with the value of the developmental parameter, indicating that developmental stability varies with the value of the developmental parameter (Fig. 4). In particular, there will be genetic variation for developmental stability only if the slope changes sufficiently between the genotypes of a locus affecting the developmental parameter. This change of slope implies that the mapping function is nonlinear and therefore there also will inevitably be an appreciable degree of dominance for the trait value.[12] This link of dominance for the trait and genetic variation for developmental instability should be a fairly general phenomenon, occurring regardless of the specific mechanisms that generate the nonlinear developmental mapping function.

In particular, this argument suggests that any gene with a significant degree of dominance for the trait value will also have an effect on developmental instability. As dominance is widespread among the genes that have effects on phenotypic traits, it follows that numerous loci will also affect developmental instability. Therefore, there is no reason to assume a priori that the control of developmental stability and its measurable outcomes, such as fluctuating asymmetry, requires specific genes.[12]

The shape of the developmental mapping function has a substantial influence on the genetic effects of a locus on developmental stability. With partial dominance for the trait, the mapping function often is a "diminishing returns" curve, initially increasing steeply and eventually approaching an asymptote, its slope is steep initially and decreases gradually towards zero. Accordingly, the effects on developmental instability will have a substantial additive component with a smaller value for the allele that is partially dominant for the trait (i.e., the a value for developmental instability and d values for the trait will have opposite signs). In contrast, a mapping function that is sigmoidal with an inflection point near the heterozygous genotype will produce overdominance for developmental instability, because the slope at the position of the heterozygote is greater than that for either homozygote. And finally, loci with overdominance or underdominance of the trait will show underdominance for developmental instability, because the slope for the mapping function is minimal near the heterozygote (Fig. 2). Some caution is necessary, however, because the patterns of dominance for developmental instability depend on the precise shape of the mapping function of the respective trait. Moreover, in nonlinear developmental systems involving multiple loci, these patterns of dominance can be modulated by the epistatic effects of other loci, so that a precise prediction of the effects of individual loci is difficult.[12]

The results of empirical studies are consistent with these patterns, although the evidence is based on only a few study systems. A classical example is the case of increased asymmetry in Australian sheep blowflies carrying mutated genes conferring insecticide resistance, which can be reduced to normal asymmetry levels by a mutant allele of a single modifier locus.[37,38] Both the resistance and modifier genes are dominant in their effects on asymmetry.[38] Searches for quantitative trait loci (QTLs) for fluctuating asymmetry in mouse mandibles and skulls found relatively high degrees of dominance.[39,40] A study of two-locus epistasis among QTLs for fluctuating asymmetry of mandible size in the mouse revealed numerous epistatic interactions and a tendency for dominance effects to be involved in these effects (additive-by-additive epistasis was relatively rare).[41] All these studies consistently underscore the importance of nonlinear effects for the genetic architecture of developmental stability.

Evolution of Developmental Instability

Developmental stability may be under selection, even though there has been considerable debate on the relationship between fitness and instability measures such as fluctuating asymmetry,[20,42-44] and if there is genetic variation in a population, there will be he potential for an evolutionary response. Two simulation studies have investigated the evolutionary dynamics of developmental instability under selection and have produced results that emphasize the importance of nonadditive genetic variation in this context.[12,45]

Klingenberg and Nijhout[12] used a nonlinear model of a diffusion-threshold process controlled by six biallelic loci to define the trait values,[25] and small random deviations were added to the parameters of the model to simulate developmental noise in this system. Trait asymmetry was computed as the difference between two values generated with the same genetic contribution, but with different random values for the contribution of developmental noise. In this model, the genotype determined the developmental mapping function for each individual and thereby mediated the response to developmental noise, but there were no loci that controlled developmental stability directly and exclusively. For each generation, a population of 10,000 offspring was generated, from which 2,000 individuals were chosen to produce the following generation.[12] Whereas selection for the trait value led to relatively quick fixation of the favored alleles, selection of fluctuating asymmetry produced a substantially slower response. Indeed, the rate of change for fluctuating asymmetry was faster when selection was on the trait value, which was correlated genetically with asymmetry, than under direct selection for fluctuating asymmetry itself. There also was a difference between selection for increased and decreased values of the trait and, even more so, between selection for increased and decreased asymmetry. The response to downward selection was slower than upward selection. Under selection for decreased fluctuating asymmetry, the evolutionary change came to a halt after approximately 40 generations.

Nonadditive genetic effects played a substantial role in shaping the evolutionary dynamics in these simulated populations.[12] There was substantial epistasis among loci, through which the alleles of each locus tended to reinforce the effects of alleles at other loci, resulting in larger allelic effects in genetic backgrounds consisting mostly of alleles favoring increased trait values. Moreover, for most population compositions, there was a positive genetic correlation between the trait value and fluctuating asymmetry. Together, these properties of the genetic system led to the unequal responses to upward and downward selection. Under downward selection, epistatic interactions tended to erode the genetic variance by decreasing the allelic effects as alleles favoring small trait values became more abundant in the genetic background. In particular, downward selection on fluctuating asymmetry virtually eliminated all additive genetic variance in the population, so that there was no further response after approximately 35 generations.[12] The analyses suggest that the difference in selection response to upward and downward selection is largely due to the effects of epistasis. In addition, selection on fluctuating asymmetry is inherently less effective than selection on trait values because of the considerable random component of variation that makes an individual's observed asymmetry a poor predictor of the underlying degree of its underlying developmental stability.[46-49]

A different model was used by Fuller and Houle[45] in their simulation study of selection on fluctuating asymmetry. They assumed a genetic model for developmental instability that did not have any direct relation to the inheritance of the trait value.[45,50] Developmental instability was implemented as a variable with a lognormal distribution, which determined the variability of the trait around its expected value. Fluctuating asymmetry for each individual was computed as the difference between two values drawn from a normal distribution with this variance. Selection was simulated for one generation, and the results were compared for simulations with different heritabilities and amounts of variation for developmental instability. The main result of the simulation was that selection was more effective for increased than for decreased asymmetry. In this model, this was due to the fact that that high asymmetry values are more informative, because they occur almost exclusively in individuals with a high degree of developmental instability, than low asymmetry values, which occur for all levels of developmental instability.[45] This result is consistent with the findings of the previous simulation of selection,[12] but adds an additional factor that contributes to the observed difference in the selection response to upward and downward selection. It is unclear, however, which of these factors is responsible for the failure to reduce fluctuating asymmetry in artificial selection experiments.[51]

Origins of Developmental Noise

Developmental systems are inherently "noisy" because they are not completely deterministic systems, but subject to random fluctuations.[52] These fluctuations originate from the molecular and cellular processes that constitute the system, and are therefore of substantial biological interest themselves. Rapid progress in the understanding of stochastic variation in gene expression, signal transduction, and other cellular processes has been made in prokaryotic and eukaryotic systems.[21,22,53-55] These random processes can contribute to variability of developmental processes and produce observable morphological variation.[1]

A number of studies have investigated stochastic variation in gene expression,[22,54,55] which may be taken as a proxy for developmental variation in general because of the fundamental importance of gene expression in developmental processes. Both transcription and translation stages have substantial influence on variability of gene expression. Transcriptional activity is related to the variability of gene product levels in variable ways in bacteria[22,54] and in a nonlinear manner with a maximum at intermediate activity in yeast,[55] presumably reflecting the differences in the transcriptional machinery between prokaryotes and eukaryotes. The variability of gene product levels has been found to be related to translational rate in a linear manner in both bacteria[54] and yeast,[55] suggesting that high efficiency of translation can amplify variation passed on from the transcriptional level.

Transcriptional processes are inherently variable because they tend to occur in bursts of activity interspersed with inactive phases.[56-59] A number of studies suggests that transcriptional regulation is achieved primarily by changing the probability of the "on" versus "off" states of the gene, rather than by adjusting the rate of transcription once the gene is activated.[56,57,60] Intermediate levels of transcription are achieved by alteration between the activated and inactive states, and therefore inevitably generate some variation in the mRNA and protein levels over time.[61] This mechanism is consistent with the finding that the variability of gene product levels tends to be greatest at intermediate levels of transcriptional activity.[55]

Cook et al[61] simulated the dynamics of stochastic gene expression in a model considering transcription, translation, and the decay of mRNA and protein product. The level of gene product is maintained at a constant average level by alternating phases of active transcription from the gene, during which the product levels will increase, with inactive phases when product levels will decrease because of the decay of mRNA and protein.[61] The precise dynamics of these processes depend on a multitude of cellular processes, but they can be summarized by the effects of the probability of the active versus inactive transcriptional state and the durations of these bouts of activity relative to the half life of the gene products.[61] If switching is fast, there will be little time for changes in gene product concentrations before the next change of transcriptional state and the ensuing reversal in the direction of change. In contrast, if switching is slow, levels will increase appreciably during the active phases and drop during the inactive phases. Therefore slow transcriptional switching results in an increased variation of gene product levels.[61]

The same model also makes predictions about the effect of gene dosage on the variability of gene product levels. Increasing the copy number of a gene has a similar effect as faster switching rate, because the chance that the different copies are active or inactive in synchrony decreases with increasing copy number. As a consequence, the variability of product levels relative to the average will decrease with increasing gene copy number.[61] Gene duplication can therefore have a stabilizing influence on developmental processes, as long as the paralogous genes are functionally equivalent.[62] In contrast, the loss of one of the two copies of a gene in a diploid organism may be associated with a substantial increase of the random fluctuations of gene product levels, which in turn may have serious consequences for the organism.[61] Indeed, this model and some extended versions of it were used by Cook et al[61] to explain the etiology of some haploinsufficiency conditions.

An experimental test of this model has been undertaken in a study of cell morphology of cultured melanocytes with different genotypes for the NF1 tumor suppressor gene.[63] The cell features studied were the number, lengths, and arrangement of the dendrites, which are long protrusions of these cells. Cells from skin samples taken from patients carrying mutations that inactivated one copy of the *NF1* gene, and where NF1 protein levels were reduced to approximately 50%, were compared to cells with two functional copies of the gene. The cells carrying the mutation showed increased variation in the number of dendrites and in the angles between dendrites, but there was no significant difference for dendrite length. Thus for two of the three cell characteristics considered, the predictions of the model by Cook et al[61] appear to hold.

These explanations of the molecular origins of stochastic variation in cellular processes are compatible, on the whole, with the approach based on nonlinear developmental mapping functions. The two approaches aim to explain the same phenomena, but are complementary in that the molecular models focus on specific mechanisms that generate random variation, whereas the framework of mapping functions is more phenomenological and concentrates on the transmission and expression of this variation in complex developmental systems. In both contexts, nonadditive effects play a substantial role via the consequences of gene dosage and through the nonlinear nature of developmental mapping functions.

Perspective: The Challenges Ahead

The Challenge of Multidimensional Phenotypes

The discussion in this chapter so far has focused entirely on the situation where the phenotype of interest is a scalar quantity, such as a length measurement, a count of repeated structural elements, or a scalar index representing a more complex aspect of morphology. Many phenotypes, however, are inherently multidimensional and cannot be easily reduced to a scalar index without severe loss of information. An example of such an inherently multidimensional phenotype is the shape of organisms or their parts, which can be quantified as a configuration of morphological landmark points[64,65] or the outline of the structure of interest.[66] In either case, the analysis of shape relies on the methods of multivariate statistics to extract relevant features of shape variation. Methods for the analysis of fluctuating asymmetry of shape are available and have been used in a variety of contexts.[67,68]

The developmental mapping functions for multidimensional phenotypes are more difficult to visualize than for scalar phenotypes, because they are not simple functions or response surfaces. The mapping function for a single input variable can be depicted as a trajectory in the multidimensional space, that is, the path on which the phenotype moves in response to the input value. This trajectory may be curved in the multidimensional space depending on the degree of nonlinearity of the developmental system. For two parameters, the mapping function is a surface, and for three parameters, it is a volume. As the number of parameters increases, this approach becomes more and more cumbersome because the mapping functions cannot be visualized any more. It is possible, however, to derive some predictions for the statistical behavior of the variation, which can be compared to the outcome of experimental treatments or comparisons of variation at different levels.[4,69,70]

Multivariate analyses of fluctuating asymmetry and variation among individuals extract patterns of variation that correspond to particular directions in phenotypic space that can provide information about the developmental system. In particular, the patterns of fluctuating asymmetry, even though they originate from random noise in developmental processes, can provide information about the developmental origins of integration among traits and other questions.[71] In the context of the present discussion on developmental mapping, a question of particular interest is the dimensionality of morphological variation. Because stochastic variation occurs in most developmental processes to some extent, the variation should be distributed over all dimensions of the parameter space, unlike genetic or environmental variation, where the available variation depends critically on the experimental design or study population used. If asymmetry variation is concentrated in only relatively few dimensions of the phenotypic space, as has been observed in several studies,[67,69,70,72] this indicates that the developmental system is channeling variation into a subset of the possible patterns. These patterns should correspond to those directions in morphological space for which the developmental mapping function is particularly steep. Therefore, it is possible to make some limited inferences about the developmental mapping function even though the relevant developmental parameters are not accessible for most empirical studies.

Directions for Future Research

This chapter has reviewed some of the theory and empirical evidence for the role of nonlinear developmental mapping, as a factor influencing the stability of developmental systems. The link between dominance of traits and genetic variation of developmental stability for those traits has been established theoretically, but has only begun to be studied empirically. A number of new approaches have been used in recent studies, which have produced an understanding of developmental noise arising within these systems and its transmission to observable phenotypic traits.

A major challenge will be to identify the mechanisms involved in developmental buffering. The example of the chaperone protein Hsp90, for which a clear role in stabilizing developmental processes has been established,[8-10] currently appears unique as a specific buffering mechanism. This mechanism is different from most of the ideas discussed in this chapter because Hsp90 is a buffering mechanism that reduces variability for many morphological traits, apparently without affecting the average trait values.[8] It is unclear, however, whether there are other similar mechanisms and whether mechanisms other than those involving chaperone molecules also have similar buffering functions. Developmental buffering phenomena are not always attributable to specific genes, but they may also result from the organization of gene regulatory networks, whose multiple feedback interactions may provide an inherent degree of stability and facilitate the evolution of developmental robustness.[11,73,74] Again, the nonlinearity of the relationships in these networks is a crucial factor determining the behavior of these systems.

Another important task will be to link this mechanistic information on developmental stability to the phenotypes affected. A significant first step has been made by linkage analysis for fluctuating asymmetry of various kinds of phenotypes.[39-41,75] The reasoning that nonlinear developmental mapping is a primary determinant of developmental instability predicts that the same genes should affect traits and fluctuating asymmetry, but the available empirical data do not allow to assess this association conclusively.[12,76] A full understanding of the link between the mechanisms conferring developmental stability and its phenotypic expression will require much additional, innovative research.

References

1. Klingenberg CP. A developmental perspective on developmental instability: Theory, models and mechanisms. In: Polak M, ed. Developmental instability: Causes and consequences. New York: Oxford University Press, 2003:14-34.
2. Waddington CH. Canalization of development and the inheritance of acquired characters. Nature 1942; 150:563-565.
3. Scharloo W. Canalization: Genetic and developmental aspects. Ann Rev Ecol Syst 1991; 22:65-93.
4. Debat V, Alibert P, David P et al. Independence between developmental stability and canalization in the skull of the house mouse. Proc Roy Soc Lond B Biol Sci 2000; 267:423-430.
5. Debat V, David P. Mapping phenotypes: Canalization, plasticity and developmental stability. Trends Ecol Evol 2001; 16:555-561.
6. Hallgrímsson B, Willmore K, Hall BK. Canalization, developmental stability, and morphological integration in primate limbs. Yearb Phys Anthropol 2002; 45:131-158.
7. Meiklejohn CD, Hartl DL. A single mode of canalization. Trends Ecol Evol 2002; 17:468-473.
8. Rutherford SL, Lindquist S. Hsp90 as a capacitor for morphological evolution. Nature 1998; 396:336-342.
9. Queitsch C, Sangster TA, Lindquist S. Hsp90 as a capacitor of phenotypic variation. Nature 2002; 417:618-624.
10. Rutherford SL. Between genotype and phenotype: Protein chaperones and evolvability. Nat Rev Genet 2003; 4:263-274.
11. von Dassow G, Meir E, Munro EM et al. The segment polarity network is a robust developmental module. Nature 2000; 406:188-192.
12. Klingenberg CP, Nijhout HF. Genetics of fluctuating asymmetry: A developmental model of developmental instability. Evolution 1999; 53:358-375.
13. West-Eberhard MJ. Adaptation: Current usages. In: Keller EF, Lloyd EA, eds. Keywords in evolutionary biology. Cambridge, MA: Harvard University Press, 1992:13-18.
14. Kacser H, Burns JA. The molecular basis of dominance. Genetics 1981; 97:639-666.
15. Orr HA. A test of Fisher's theory of dominance. Proc Natl Acad Sci USA 1991; 88:11413-11415.
16. Porteous JW. Dominance—one hundred and fifteen years after Mendel's paper. J Theor Biol 1996; 182:223-232.

17. Mayo O, Bürger R. The evolution of dominance: A theory whose time has passed? Biol Rev 1997; 72:97-110.
18. Bourguet D. The evolution of dominance. Heredity 1999; 83:1-4.
19. Palmer AR, Strobeck C. Fluctuating asymmetry: Measurement, analysis, patterns. Ann Rev Ecol Syst 1986; 17:391-421.
20. Møller AP, Swaddle JP. Asymmetry, developmental stability, and evolution. Oxford: Oxford University Press, 1997.
21. Swain PS, Elowitz MB, Siggia ED. Intrinsic and extrinsic contributions to stochasticity in gene expression. Proc Natl Acad Sci USA 2002; 99:12795-12800.
22. Elowitz MB, Levine AJ, Siggia ED et al. Stochastic gene expression in a single cell. Science 2002; 297:1183-1186.
23. Kell DB. Genotype-phenotype mapping: Genes as computer programs. Trends Genet 2002; 18:555-559.
24. Wagner GP. Homologues, natural kinds and the evolution of modularity. Amer Zool 1996; 36:36-43.
25. Nijhout HF, Paulsen SM. Developmental models and polygenic characters. Am Nat 1997; 149:394-405.
26. Gilchrist MA, Nijhout HF. Nonlinear developmental processes as sources of dominance. Genetics 2001; 159:423-432.
27. Veitia RA. Exploring the etiology of haploinsufficiency. BioEssays 2002; 24:175-184.
28. Falconer DS, Mackay TFC. Introduction to quantitative genetics. 4th ed. Essex: Longman, 1996.
29. Lynch M, Walsh B. Genetics and analysis of quantitative traits. Sunderland, MA: Sinauer, 1998.
30. Phillips PC. The language of gene interaction. Genetics 1998; 149:1167-1171.
31. Davidson EH. Genomic regulatory systems: Development and evolution. San Diego: Academic Press, 2001.
32. Wilkins AS. The evolution of developmental pathways. Sunderland, MA: Sinauer Associates, 2002.
33. Templeton AR. Epistasis and complex traits. In: Wolf JB, Brodie III ED, Wade MJ, eds. Epistasis and the evolutionary process. New York: Oxford University Press, 2000:41-57.
34. Cheverud JM, Routman EJ. Epistasis and its contribution to genetic variance components. Genetics 1995; 139:1455-1461.
35. Cheverud JM, Routman EJ. Epistasis as a source of increased additive genetic variance at population bottlenecks. Evolution 1996; 50:1042-1051.
36. Routman EJ, Cheverud JM. Gene effects on a quantitative trait: Two-locus epistatic effects measured at microsatellite markers and at estimated QTL. Evolution 1997; 51:1654-1662.
37. McKenzie JA, Clarke GM. Diazinon resistance, fluctuating asymmetry and fitness in the Australian sheep blowfly, Lucilia cuprina. Genetics 1988; 120:213-220.
38. McKenzie JA. The analysis of the asymmetry phenotype: Single genes and the environment. In: Polak M, ed. Developmental instability: Causes and consequences. New York: Oxford University Press, 2003:135-141.
39. Leamy LJ, Routman EJ, Cheverud JM. A search for quantitative trait loci affecting asymmetry of mandibular characters in mice. Evolution 1997; 51:957-969.
40. Leamy LJ, Routman EJ, Cheverud JM. Quantitative trait loci for fluctuating asymmetry of discrete skeletal characters in mice. Heredity 1998; 80:509-518.
41. Leamy LJ, Routman EJ, Cheverud JM. An epistatic genetic basis for fluctuating asymmetry of mandible size in mice. Evolution 2002; 56:642-653.
42. Møller AP. Developmental stability and fitness: A review. Am Nat 1997; 149:916-932.
43. Clarke GM. Developmental stability-fitness relationships in animals: Some theoretical considerations. In: Polak M, ed. Developmental stability: Causes and consequences. New York: Oxford University Press, 2003:187-195.
44. Lens L, van Dongen S, Kark S et al. Fluctuating asymmetry as an indicator of fitness: Can we bridge the gap between studies? Biol Rev 2002; 77:27-38.
45. Fuller RC, Houle D. Detecting genetic variation in developmental instability by artificial selection on fluctuating asymmetry. J Evol Biol 2002; 15:954-960.

46. Whitlock M. The heritability of fluctuating asymmetry and the genetic control of developmental stability. Proc Roy Soc Lond B Biol Sci 1996; 263:849-854.
47. Whitlock M. The repeatability of fluctuating asymmetry: A revision and extension. Proc Roy Soc Lond B Biol Sci 1998; 265:1429-1431.
48. Van Dongen S. How repeatable is the estimation of developmental stability by fluctuating asymmetry? Proc Roy Soc Lond B Biol Sci 1998; 265:1423-1427.
49. van Dongen S, Lens L, Molenberghs G. Recent developments and shortcomings in the analysis of individual asymmetry: A review and introduction of a Bayesian statistical approach. In: Polak M, ed. Developmental instability: Causes and consequences. New York: Oxford University Press, 2003:320-342.
50. Fuller RC, Houle D. Inheritance of developmental instability. In: Polak M, ed. Developmental instability: Causes and consequences. New York: Oxford University Press, 2003:157-183.
51. Breuker CJ, Brakefield PM. Lack of response to selection for lower fluctuating asymmetry of mutant eyespots in the butterfly Bicyclus anynana. Heredity 2003; 91:17-27.
52. McAdams HH, Arkin A. It's a noisy business! Genetic regulation at the nanomolecular scale. Trends Genet 1999; 15:65-69.
53. Smolen P, Baxter DA, Byrne JH. Modeling transcriptional control in gene networks—Methods, recent results, and future directions. Bull Math Biol 2000; 62:247-292.
54. Ozbudak EM, Thattai M, Kurtser I et al. Regulation of noise in the expression of a single gene. Nat Genet 2002; 31:69-73.
55. Blake WJ, Kaern M, Cantor CR et al. Noise in eukaryotic gene expression. Nature 2003; 422:633-637.
56. Hume DA. Probability in transcriptional regulation and its implications for leukocyte differentiation and inducible gene expression. Blood 2000; 96:2323-2328.
57. Fiering S, Whitelaw E, Martin DIK. To be or not to be active: The stochastic nature of enhancer action. BioEssays 2000; 22:381-387.
58. Ko MSH. Induction mechanism of a single gene molecule: Stochastic or deterministic? BioEssays 1992; 14:341-346.
59. Ross IL, Browne CM, Hume DA. Transcription of individual genes in eukaryotic cells occurs randomly and infrequently. Immun Cell Biol 1994; 72:177-185.
60. Walters MC, Fiering S, Eidemiller J et al. Enhancers increase the probability but not the level of gene expression. Proc Natl Acad Sci USA 1995; 92:7125-7129.
61. Cook DL, Gerber AN, Tapscott SJ. Modeling stochastic gene expression: Implications for haploinsufficiency. Proc Natl Acad Sci USA 1998; 95:15641-15646.
62. Wilkins AS. Canalization: A molecular genetic perspective. BioEssays 1997; 19:257-262.
63. Kemkemer R, Schrank S, Vogel W et al. Increased noise as an effect of haploinsufficiency of the tumor-suppressor gene neurofibromatosis type 1 in vitro. Proc Natl Acad Sci USA 2002; 99:13783-13788.
64. Bookstein FL. Morphometric tools for landmark data: Geometry and biology. Cambridge: Cambridge University Press, 1991.
65. Dryden IL, Mardia KV. Statistical analysis of shape. Chichester: Wiley, 1998.
66. Lestrel PE, ed. Fourier descriptors and their applications in biology. Cambridge: Cambridge University Press, 1997.
67. Klingenberg CP, McIntyre GS. Geometric morphometrics of developmental instability: Analyzing patterns of fluctuating asymmetry with Procrustes methods. Evolution 1998; 52(5):1363-1375.
68. Klingenberg CP, Barluenga M, Meyer A. Shape analysis of symmetric structures: Quantifying variation among individuals and asymmetry. Evolution 2002; 56:1909-1920.
69. Klingenberg CP, Zaklan SD. Morphological integration between developmental compartments in the Drosophila wing. Evolution 2000; 54:1273-1285.
70. Klingenberg CP, Mebus K, Auffray J-C. Developmental integration in a complex morphological structure: How distinct are the modules in the mouse mandible? Evol Dev 2003; 5:522-531.
71. Klingenberg CP. Developmental instability as a research tool: Using patterns of fluctuating asymmetry to infer the developmental origins of morphological integration. In: Polak M, ed. Developmental Instability: Causes and Consequences. New York: Oxford University Press, 2003:427-442.

72. Klingenberg CP, Badyaev AV, Sowry SM et al. Inferring developmental modularity from morphological integration: Analysis of individual variation and asymmetry in bumblebee wings. Am Nat 2001; 157:11-23.
73. Houchmandzadeh B, Wieschaus E, Leibler S. Establishment of developmental precision and proportions in the early Drosophila embryo. Nature 2002; 415:798-802.
74. von Dassow G, Odell GM. Design and constraints of the Drosophila segment polarity module: Robust spatial patterning emerges from intertwined cell state switches. J Exp Zool (Mol Dev Evol) 2002; 294:179-215.
75. Klingenberg CP, Leamy LJ, Routman EJ et al. Genetic architecture of mandible shape in mice: Effects of quantitative trait loci analyzed by geometric morphometrics. Genetics 2001; 157:785-802.
76. Leamy LJ. Dominance, epistasis, and fluctuating asymmetry. In: Polak M, ed. Developmental instability: Causes and consequences. New York: Oxford University Press, 2003:142-156.

Chapter 4

Phenotype and Stochastic Gene Expression:

Can the Noise Cause Haploinsufficiency?

Anthony N. Gerber, D.L. Cook and S.J. Tapscott

Abstract

Growing evidence suggests that genes are not expressed at steady state levels, but instead, gene transcription has a stochastic component. We review recent literature concerning stochastic gene expression and discuss our model in which allele copy number influenced the probability that the product level of a gene will fall below a critical level in an individual cell over time. In this model, loss of a single allele caused auto-regulated genes to be permanently inactivated. Implications for haploinsufficent phenotypes and redundant gene families are discussed. Specific genes that have been associated with haploinsufficiency are considered in the context of emerging studies and models of stochastic gene expression.

Deterministic verses Stochastic Gene Expression

Haploinsufficiency refers to a phenotype associated with the functional loss of a single allele of a gene in a diploid organism. Traditionally, haploinsufficiency has been considered to result from a 50% reduction in gene expression associated with such a loss; this model assumes that normal gene expression is at relatively stable levels from both alleles, and that expression levels in the haploid state are not increased from the remaining allele (dosage compensation). Several syndromes have been characterized molecularly that appear to fit this basic model. For example, the COL5A1 gene, which encodes the proα1(V) chain of type V collagen, exhibited heterozygous loss of function mutations in a group of patients with the Ehler-Danlos syndrome.[1] This resulted in a decrease in proα1(V) RNA levels, with expression only observed from the nonmutant allele. Since precise stochiometric relationships of collagen precursor protein levels are required for the correct assembly of mature collagen,[2] the reduction in proα1(V) levels associated with mutation in COL5A1 results in an Ehler-Danlos phenotype.

While this example demonstrates that phenotypes can be associated with a 50% reduction in the steady state level of a gene product, it is likely that some haploinsufficiency syndromes are caused by a different mechanism: stochastic delays or interruptions of gene expression. The amount of gene product in a cell is determined by the integration of a series of quantal events: The assembly of a transcription complex at the gene promoter, the release of the polymerase holoenzyme complex to generate an RNA transcript, the translation of the mRNA into protein by the ribosome, and the degradation of the protein, to name just a few. The rate that each of these events will occur depends, in part, on the relative concentration of the component parts and their probability of forming a functional complex. For example, the probability that a transcription factor will bind its regulatory element increases with the concentration of the factor until a saturating level is achieved.[3,4] In this regard, gene expression can be viewed as a series of probabilistic reactions, or as a series of stochastic events.

In the case of gene transcription, at saturating amounts of inducer the expression might be immediate and continuous; whereas at intermediate or low levels of inducer the onset of transcription might be delayed and intermittent. In a model system of steroid induced gene expression using the mouse mammary tumor virus regulatory region (a dexamethasone responsive promoter) driving the β-galactosidase reporter gene, exposure to increasing amounts of dexamethasone increased the amount of β-galactosidase by increasing the number of cells expressing β-galactosidase rather than causing a graded increase in each expressing cell, indicating that the increasing concentration of the inducer increased the probability that the target gene would be expressed.[5] Also, macrophages exposed to different levels of the inducing agent lipopolysaccharide showed correspondingly increasing mRNA levels of the response gene PAI-2. Again, this was due to an increasing percentage of cells exhibiting high levels of PAI-2 expression, rather than all the cells having a graded increase in a continuum of response.[6] Similarly, in a transgenic mouse model, a globin enhancer increased the number of erythroid cells that expressed beta galactosidase, but did not change the expression level in individual cells.[7] Together, these studies and numerous others[8-11] are consistent with the conclusion that the graded response of genes to inducing agents is achieved by altering the probability that a gene will be expressed, rather than the rate at which transcription proceeds after initiation.

If inducers act to increase the probability that a gene is expressed, then at intermediate levels of inducers there must be a dynamic equilibrium between active expression and de-activation of the gene. This was demonstrated in prokaryotes using bacterial lines with two distinguishable alleles of green fluorescent protein, cyan (CFP) and yellow (YFP), driven by an IPTG inducible promoter. These fluorescent reporters allowed both real time and continuous assessment of protein expression since fluorescence could be assessed in living cells. At intermediate levels of the IPTG inducer, relatively random and intermittent expression of each allele was observed. The authors concluded that gene expression is stochastic, with variability in expression between alleles within the same cells.[12]

The scant amount of data available in eukaryotic cells also suggests a dynamic equilibrium between the "on" and "off" states of gene expression. One study utilized in situ hybridization to the nascent transcripts of the carbamoylphosphate synthase RNA in glucocorticoid treated hepatocytes.[13] Expression was both random and intermittent, indicating stochastic variation between active and inactive states at the level of gene expression. Similar stochastic transitions might be inferred from patterns of gene expression in skeletal muscle. Mature muscle cells have many nuclei within a common cytoplasm, yet, active transcription of a number of muscle structural genes was only observed in a random subset of nuclei within a fiber, as assessed by in situ hybridization.[14] Since muscle fibers are structurally homogeneous across much of their length, it was concluded that stochastic mechanisms must govern gene expression in the myofiber.

There are few direct measurements of the kinetics of gene expression. An analysis of the, δ, γ and β globin genes indicated that expression of each gene was intermittent and estimated dwell times of gene expression ranged from 4 min for the ε gene to 45-80 min for the β gene during a developmental transition.[15] There is also evidence that some transcription factor complexes actively cycle between bound and unbound states. In one study, the binding kinetics of the estrogen receptor to the regulatory region of several estrogen responsive genes was determined.[16] Receptor binding peaked 45 minutes after exposure to a saturating amount of ligand, then binding diminished, then binding peaked again at 135 minutes after the initial stimulation. Although binding is a surrogate marker for gene expression, it is reasonable to assume that the estrogen gene cycles between on and off states at intervals on the order of an hour. The rapid de-activation kinetics of the estrogen responsive genes might be necessary to achieve a graded response to intermediate levels of inducers. In this model, an increasing amount of

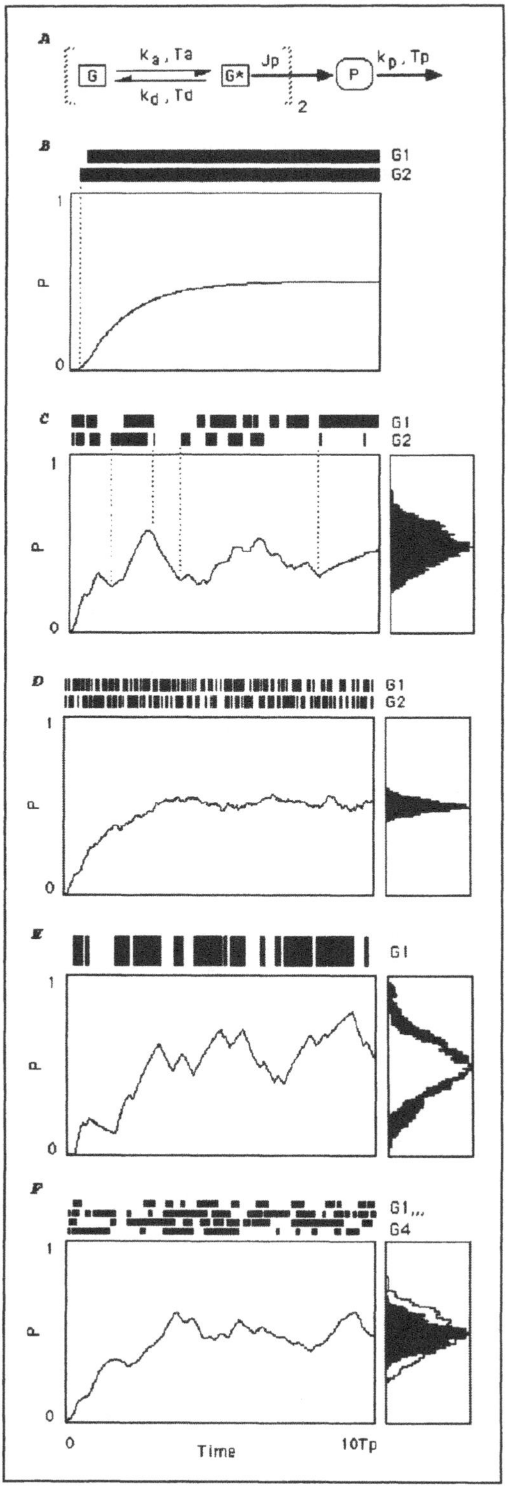

Figure 1. A stochastic model of gene expression. A) This minimal model of stochastic gene expression kinetics consists of a pool of product, P, and two identical genes (indicated by brackets and the subscript 2). P is degraded according to first-order kinetics with a half time, Tp. (For a first-order process, the rate constant corresponding to a half time, T, is k = log2/T.) Each gene functions independently of the other and can be inactive (G) or active (G*). Each active gene expresses P at a rate Jp (in nM/s, for instance) so that, if two genes are active, then the net rate of P-synthesis is twice Jp. Genes switch spontaneously between active and inactive state according to first-order kinetics with an activation half time of Ta and a deactivation half time of Td (and corresponding rate constants, k_a and k_d). B) Both genes (G1 and G2) are initially inactive and then are allowed to activate (as indicated by the black bars at the top) independently and randomly. In this simulation, the deactivation rate (k_d) was 0 and expression was stable after activation. C) The deactivation half time was set to match the activation half time used in A. To the right, an amplitude histogram shows the dispersion, or variance, of the product accumulated over time, i.e., the expression noise. D) Activation and deactivation kinetics that were 10-fold faster than in A created such brief active and inactive periods that the product level changed only slightly. This markedly reduced product variance. E) When a single gene synthesized product with the same kinetics as in C, the dispersion of expression noise increased as seen by comparing the black histogram with the white histogram taken from C. F) When the same overall synthesis rate was distributed among four genes, the expression noise was reduced compared with that of two genes (C). This figure and portions of the legend were published previously as referenced in the text.

ligand would increase the probability of turning the gene on and thereby increase the average amount of time the gene was transcribed because of the constant off-rate.

Several groups have incorporated stochastic parameters of gene expression in simulations of biological systems. Ko developed a simple model of stochastic gene expression based on experimental observation of the induction of gene expression by steroid hormone.[17] Groups have also modeled extrinsic and intrinsic noise in simple genetic circuits,[18] the lysis/lysogeny decision in phage lambda,[19] and the effects of kinetics on stability and escape time,[20] among others (for more detailed mathematical review of modeling concepts and other available stochastic expression models see refs. 21-23). In general, insufficient quantitative data regarding key parameters that govern gene expression has led to qualitative rather than quantitative conclusions from most models.

Haploinsufficiency in a Model of Stochastic Gene Expression

We modeled the qualitative effects of gene copy number on threshold events associated with gene expression falling below a given level. In our previously published model,[24] gene expression switched randomly between active and inactive states with first order kinetics. When the gene deactivation rate constant was greater than zero, gene expression was intermittent (Fig. 1C). This resulted in fluctuations around a mean expression level. Increasing gene copy number reduced the amplitude of the fluctuations (Fig. 1F); conversely decreasing allele number from two to one, while keeping average product synthesis constant (i.e., allowing dosage compensation), increased fluctuation amplitude (Fig. 1E). Thus, reduction in gene copy number, even in the presence of dosage compensation, can result in fluctuations of product quantity to lower levels than would be observed in wild type cells. If such fluctuations are associated with a deleterious effect on the cell, a phenotype would be expected to result. Moreover, in our model, if the average rate of product synthesis was allowed to drop by 50% with the loss of one allele (i.e., no dosage compensation), transient fluctuations to significantly below the 50% product level would occur.

To illustrate the potential effects of gene expression fluctuations below a threshold level, we modeled a simple auto-regulated genetic network in which the gene product stimulated its own expression. A certain level of product was required in this model to maintain a nonzero probability of gene expression, with decreases in product quantity below this level resulting in irreversible termination of gene expression. In our simulations, the majority of decreases in product associated with stochastic deactivation of expression did not result in product levels dropping below the threshold level. Occasional prolonged lapses in expression occurred, however, which resulted in permanent inactivation of expression (Fig. 2). Times prior to deactivation in our one-gene model followed an exponential distribution; average survival time could range from minutes to years depending on the relative values of the expression kinetics, half-lives, and the threshold value. If threshold events for a specific gene are associated with effects on cell survival or function, inactivation of a single allele in a diploid organism could have phenotypic manifestations at any point during the lifetime of an organism, ranging from developmental defects to late onset degenerative diseases. Moreover, the probabilistic behavior that governs the accumulation of events in this model implies that phenotypes and onset age for haploinsufficient syndromes could vary widely between individuals, as is observed clinically.

Many variables could affect gene product levels including the half-lives of mRNA and protein, and the efficiency of transcriptional elongation and translation. In our model, we collapsed production and degradation kinetics into a single product pool and kept this variable constant to analyze the properties of expression kinetics in isolation. A bacterial model using a GFP reporter has been used to argue that translation efficiency is the major factor behind observed variations in GFP levels.[25] However, recent eukaryotic work using a stochastic expression model in yeast[26] showed that stochastic variations in mRNA levels were the major determinant

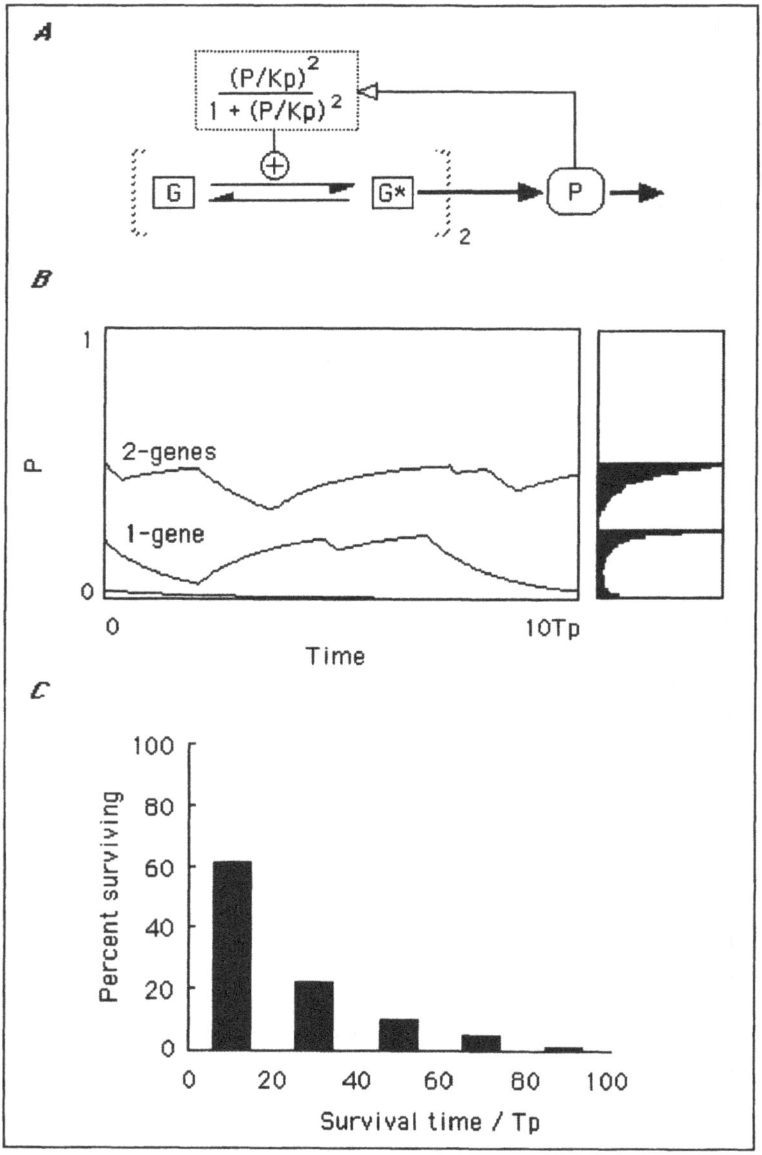

Figure 2. Maintenance of auto-regulated gene expression is sensitive to copy number in a stochastic manner. A) The model of (Fig. 1) was modified to simulate simultaneous binding (with affinity Kp) of two P molecules to a receptor that activates gene expression, and the expression deactivation rate was decreased 10-fold (Td = 10Tp/4) to simulate a relatively stable gene with infrequent expression lapses. The expression rate, Jp, was decreased to 12.2 to maintain an average product level of ~0.5 uP. Setting Kp = 0.2 of the full-scale value of P = 1.0 established an all-or-none activation/deactivation threshold at a value P = 0.05. B) When two genes were active (upper trace and histogram), P levels were maintained well above the threshold, and the genes were expressed indefinitely. However, when one gene was inactivated, random lapses of expression allowed P levels to fall below the threshold level (lower trace continued as a second line, and lower histogram). C) Histogram of survival times for 100 trials of a one-gene model starting with active expression and an initial P of 0.25. This figure and portions of the text were previously published as referenced in the text.

of experimentally observed fluctuations of a GFP reporter. In both the bacterial and yeast systems, higher translational efficiencies were associated with higher levels of noise. This is predicted by our model and others,[18,24] since highly efficient translation would cause a more direct correlation between stochastic fluctuations in mRNA levels and protein levels, while, on the other hand, the large number of RNA molecules that are on average produced from an active gene provide a buffer against noise associated with inefficient translation. Genes that are expressed at very low levels with relatively stable expression kinetics would be a potential caveat to this notion. As technology allows more precise analysis of gene expression on the cellular level, it will become possible to dissect the relative effects of stochastic events governing transcription, translation and degradation on biological behavior.

Evolutionary Considerations

Why then would nature select for expression kinetics that makes genes vulnerable to haploinsufficient phenotypes? One important paradigm that has emerged with regard to gene regulation is the frequent use of both positive and negative control elements. It is possible that blockade of transcription by some negative regulators requires disassociation of previously active transcription complexes from DNA. In developmental transitions, therefore, the necessity of precise temporal and spatial control of gene expression may require rapidly cycling kinetics. Fast on and off kinetics may also allow for sharper responses to a graded signal.[17,24] In this regard, a growing number of signaling molecules have been identified that are associated with haploinsufficiency. For example, heterozygous deletion of the signal transduction protein, LIS-1, can cause Miller-Dieker lissencephaly syndrome,[27] and haploinsufficiency of the hedgehog receptor, patched, is associated with Gorlin's syndrome.[28] It is also possible that activation-deactivation kinetics have evolved at some genes to permit the integrity of each allele to be sampled, a process that can increase population fitness.[29]

The binary nature of many cell fate decisions controlled by transcription factors suggests a role for bi-stable systems in development. Fast expression kinetics, which have been associated with bi-stable behavior in model networks,[30] may cause certain transcription factors to be vulnerable to haploinsufficient phenotypes; additionally, genetic networks that auto-regulate are sensitive to copy number with regard to threshold events. In support of these notions, an increasing number of transcription factors have now been identified that exhibit haploinsufficiency; many of these are in families that are believed to auto-regulate. For example, mutations in GLI3, a gene that is involved in a complex self-regulatory network,[31-33] can cause Greig cephalopolysyndactyly syndrome.[34] Homeobox genes have been implicated in auto-regulation, and haploinsufficiency of the ALX4 homeobox gene is associated with skull ossification defects,[35] while MSX1 haploinsufficiency is associated with tooth agenesis.[36] Other examples of transcription factors associated with haploinsufficiency include GATA4, which has a haploinsufficient congenital heart disease phenotype,[37] SOX9 and SOX10, which are associated with campomelic dysplasia and the Waardenburg-Shah syndrome respectively;[38,39] and the thyroid transcription factor NKX2.1, which causes haploinsufficient neurological defects.[40] Heterozygous loss of function mutation of TWIST causes the SaethreChotzen syndrome; haploinsufficiency of PAX6 generates cerebral malformations and olfactory dysfunction, while mutations in PAX9 are associated with autosomal hypodontia;[41-44] indeed, the list of transcription factors associated with human haploinsufficiency syndromes is large and growing rapidly. Gene targeting experiments have also identified a large number of transcription factors and signaling molecules with haploinsufficient phenotypes in mice.

If rapid stochastic expression kinetics were selected to allow execution of certain developmental decisions, one would predict that mechanisms to protect against associated haploinsufficient phenotypes would have developed. One such mechanism may be gene duplication and network redundancy. In our model, increasing gene copy number from two to four reduced fluctuations in expression levels. Similar results have been observed experimentally when lower expression noise was associated with the 2n post- replication phase of the bacterial cell cycle.[18] In higher organisms, genetic analysis has revealed a large number of genes that are at least partially redundant, some of which have been shown to exhibit complex, synthetic, trans-heterozygote phenotypes. For example, haploinsufficiency of the winged/forkhead transcription factor, FOXC2, has been associated with the Lymphadema-distichiasis syndrome in humans,[45] while in mice, homozygous null mutations of Foxc1 and the closely related Foxc2 cause multiple developmental defects including hydrocephalus and duplex kidneys. Compound Foxc1$^{+/-}$; Foxc2$^{+/-}$ heterozygotes exhibited similar rates of kidney anomalies,[46] demonstrating functional redundancy within the family, and demonstrating that loss of 50% of the alleles within a redundant family can be associated with a phenotype. Detailed analysis of compound heterozygote mice for other sets of closely related transcription factors may reveal more examples of phenotype associated with graded reductions in gene dosage.

Many haploinsufficient diseases exhibit significant variability in phenotypic severity—i.e., variable penetrance.[47,48] Variable penetrance could itself be a consequence of stochastic gene expression, since potential threshold events associated with a phenotype would occur randomly over time. It is also possible that variable penetrance of haploinsufficient diseases in genetically heterogeneous populations may be related to unlinked genetic modifiers that influence the stochastic properties of gene expression. In this regard, different strains of bacteria have been shown to exhibit different levels of expression noise.[12] In addition, a role for chromatin factors in stochastic gene expression has been suggested, which would imply that allelic variation in a variety of cellular factors affecting chromatin structure could influence stochastic expression properties.[15,49]

The regulation of expression kinetics by chromatin structure would potentially allow for dynamic changes in the stochastic properties governing expression of a given gene. For example, a gene may exhibit rapid kinetics at a developmental transition, after which the local chromatin structure may remodel and change the rate of transcriptional fluctuation. Thus, it is possible that haploinsufficiency of a given gene may cause errors at a specific developmental point, and subsequently predispose to the more gradual accumulation of post-developmental threshold events affecting cell function. Exemplifying this notion, Ptc1$^{+/-}$ mice exhibit a low rate of polydactyly during embryogenesis, while as adults, they are predisposed to develop several types of cancer at increased rates.[28,50,51] It will be of interest to correlate chromatin structure, off rates, and haploinsufficiency. Tissue specific haploinsufficient phenotypes of broadly expressed genes may be secondary to locally regulated expression kinetics.

Haploinsufficiency can be viewed as an extreme example of the contributions that two alleles of a gene can make to phenotype in a diploid organism. Just as expression kinetics may in part govern haploinsufficiency, expression kinetics may also influence how the inputs of two alleles are synthesized during the execution of relatively stereotyped developmental programs. In this regard, complex inheritance patterns may reflect not only the underlying segregation of genetic modifiers, but the expression kinetics of the involved loci. Further exploration of the relationship between haploinsufficiency and stochastic gene expression will yield insights into the general consequences of expression noise on phenotype and dominance.

References

1. Wenstrup RJ, Florer JB, Willing MC et al. COL5A1 haploinsufficiency is a common molecular mechanism underlying the classical form of EDS. Am J Hum Genet 2000; 66(6):1766-1776.
2. De Paepe A. Heritable collagen disorders: From phenotype to genotype. Verh K Acad Geneeskd Belg 1998; 60(5):463-482. discussion 482-464.
3. Boulanger PA, Yoshinaga SK, Berk AJ. DNA-binding properties and characterization of human transcription factor TFIIIC2. J Biol Chem 1987; 262(31):15098-15105.

4. Kato H, Nagamine M, Kominami R et al. Formation of the transcription initiation complex on mammalian rDNA. Mol Cell Biol 1986; 6(10):3418-3427.
5. Ko MS, Nakauchi H, Takahashi N. The dose dependence of glucocorticoid-inducible gene expression results from changes in the number of transcriptionally active templates. EMBO J 1990; 9(9):2835-2842.
6. Costelloe EO, Stacey KJ, Antalis TM et al. Regulation of the plasminogen activator inhibitor-2 (PAI-2) gene in murine macrophages. Demonstration of a novel pattern of responsiveness to bacterial endotoxin. J Leukoc Biol 1999; 66(1):172-182.
7. Sutherland HG, Martin DI, Whitelaw E. A globin enhancer acts by increasing the proportion of erythrocytes expressing a linked transgene. Mol Cell Biol 1997; 17(3):1607-1614.
8. van Roon MA, Aten JA, van Oven CH et al. The initiation of hepatocyte-specific gene expression within embryonic hepatocytes is a stochastic event. Dev Biol 1989; 136(2):508-516.
9. Fiering S, Whitelaw E, Martin DI. To be or not to be active: The stochastic nature of enhancer action. Bioessays 2000; 22(4):381-387.
10. Weintraub H. Formation of stable transcription complexes as assayed by analysis of individual templates. Proc Natl Acad Sci USA 1988; 85(16):5819-5823.
11. Siegele DA, Hu JC. Gene expression from plasmids containing the araBAD promoter at subsaturating inducer concentrations represents mixed populations. Proc Natl Acad Sci USA 1997; 94(15):8168-8172.
12. Elowitz MB, Levine AJ, Siggia ED et al. Stochastic gene expression in a single cell. Science 2002; 297(5584):1183-1186.
13. Dingemanse MA, de Boer PA, Moorman AF et al. The expression of liver-specific genes within rat embryonic hepatocytes is a discontinuous process. Differentiation 1994; 56(3):153-162.
14. Newlands S, Levitt LK, Robinson CS et al. Transcription occurs in pulses in muscle fibers. Genes Dev 1998; 12(17):2748-2758.
15. Wijgerde M, Grosveld F, Fraser P. Transcription complex stability and chromatin dynamics in vivo. Nature 1995; 377(6546):209-213.
16. Shang Y, Hu X, DiRenzo J et al. Cofactor dynamics and sufficiency in estrogen receptor-regulated transcription. Cell 2000; 103(6):843-852.
17. Ko MS. A stochastic model for gene induction. J Theor Biol 1991; 153(2):181-194.
18. Swain PS, Elowitz MB, Siggia ED. Intrinsic and extrinsic contributions to stochasticity in gene expression. Proc Natl Acad Sci USA 2002; 99(20):12795-12800.
19. Arkin A, Ross J, McAdams HH. Stochastic kinetic analysis of developmental pathway bifurcation in phage lambda-infected Escherichia coli cells. Genetics 1998; 149(4):1633-1648.
20. Kierzek AM, Zaim J, Zielenkiewicz P. The effect of transcription and translation initiation frequencies on the stochastic fluctuations in prokaryotic gene expression. J Biol Chem 2001; 276(11):8165-8172.
21. de Jong H. Modeling and simulation of genetic regulatory systems: A literature review. J Comput Biol 2002; 9(1):67-103.
22. Smolen P, Baxter DA, Byrne JH. Modeling transcriptional control in gene networks—methods, Recent results, and future directions. Bull Math Biol 2000; 62(2):247-292.
23. Hasty J, McMillen D, Isaacs F et al. Computational studies of gene regulatory networks: In numero molecular biology. Nat Rev Genet 2001; 2(4):268-279.
24. Cook DL, Gerber AN, Tapscott SJ. Modeling stochastic gene expression: Implications for haploinsufficiency. Proc Natl Acad Sci USA 1998; 95(26):15641-15646.
25. Ozbudak EM, Thattai M, Kurtser I et al. Regulation of noise in the expression of a single gene. Nat Genet 2002; 31(1):69-73.
26. Blake WJ, M KA, Cantor CR et al. Noise in eukaryotic gene expression. Nature 2003; 422(6932):633-637.
27. Reiner O, Carrozzo R, Shen Y et al. Isolation of a Miller-Dieker lissencephaly gene containing G protein beta-subunit-like repeats. Nature 1993; 364(6439):717-721.
28. Pazzaglia S, Mancuso M, Atkinson MJ et al. High incidence of medulloblastoma following X-ray-irradiation of newborn Ptc1 heterozygous mice. Oncogene 2002; 21(49):7580-7584.
29. Maley CC, Tapscott SJ. Selective Instability: Maternal Evolution of Gene Activation and Deactivation Rates. Artificial Life 2003; 9(3):in press.

30. Kepler TB, Elston TC. Stochasticity in transcriptional regulation: Origins, consequences, and mathematical representations. Biophys J 2001; 81(6):3116-3136.
31. Bailey EC, Scott MP, Johnson RL. Hedgehog signaling in animal development and human disease. Ernst Schering Res Found Workshop 2000; (29):211-235.
32. Bai CB, Joyner AL. Gli1 can rescue the in vivo function of Gli2. Development 2001; 128(24):5161-5172.
33. Park HL, Bai C, Platt KA et al. Mouse Gli1 mutants are viable but have defects in SHH signaling in combination with a Gli2 mutation. Development 2000; 127(8):1593-1605.
34. Kalff-Suske M, Wild A, Topp J et al. Point mutations throughout the GLI3 gene cause Greig cephalopolysyndactyly syndrome. Hum Mol Genet 1999; 8(9):1769-1777.
35. Mavrogiannis LA, Antonopoulou I, Baxova A et al. Haploinsufficiency of the human homeobox gene ALX4 causes skull ossification defects. Nat Genet 2001; 27(1):17-18.
36. Hu G, Vastardis H, Bendall AJ et al. Haploinsufficiency of MSX1: A mechanism for selective tooth agenesis. Mol Cell Biol 1998; 18(10):6044-6051.
37. Pehlivan T, Pober BR, Brueckner M et al. GATA4 haploinsufficiency in patients with interstitial deletion of chromosome region 8p23.1 and congenital heart disease. Am J Med Genet 1999; 83(3):201-206.
38. Cameron FJ, Hageman RM, Cooke-Yarborough C et al. A novel germ line mutation in SOX9 causes familial campomelic dysplasia and sex reversal. Hum Mol Genet Oct 1996; 5(10):1625-1630.
39. Pingault V, Bondurand N, Kuhlbrodt K et al. SOX10 mutations in patients with Waardenburg-Hirschsprung disease. Nat Genet 1998; 18(2):171-173.
40. Krude H, Schutz B, Biebermann H et al. Choreoathetosis, hypothyroidism, and pulmonary alterations due to human NKX2-1 haploinsufficiency. J Clin Invest 2002; 109(4):475-480.
41. Howard TD, Paznekas WA, Green ED et al. Mutations in TWIST, a basic helix-loop-helix transcription factor, in SaethreChotzen syndrome. Nat Genet 1997; 15(1):36-41.
42. el Ghouzzi V, Le Merrer M, Perrin-Schmitt F et al. Mutations of the TWIST gene in the SaethreChotzen syndrome. Nat Genet 1997; 15(1):42-46.
43. Sisodiya SM, Free SL, Williamson KA et al. PAX6 haploinsufficiency causes cerebral malformation and olfactory dysfunction in humans. Nat Genet 2001; 28(3):214-216.
44. Das P, Stockton DW, Bauer C et al. Haploinsufficiency of PAX9 is associated with autosomal dominant hypodontia. Hum Genet 2002; 110(4):371-376.
45. Fang J, Dagenais SL, Erickson RP et al. Mutations in FOXC2 (MFH-1), a forkhead family transcription factor, are responsible for the hereditary lymphedema-distichiasis syndrome. Am J Hum Genet 2000; 67(6):1382-1388.
46. Kume T, Deng K, Hogan BL. Murine forkhead/winged helix genes Foxc1 (Mf1) and Foxc2 (Mfh1) are required for the early organogenesis of the kidney and urinary tract. Development 2000; 127(7):1387-1395.
47. Viskochil D. Genetics of neurofibromatosis 1 and the NF1 gene. J Child Neurol 2002; 17(8):562-570.
48. Dollfus H, Biswas P, Kumaramanickavel G et al. SaethreChotzen syndrome: Notable intrafamilial phenotypic variability in a large family with Q28X TWIST mutation. Am J Med Genet 2002; 109(3):218-225.
49. Ahmad K, Henikoff S. Modulation of a transcription factor counteracts heterochromatic gene silencing in Drosophila. Cell 2001; 104(6):839-847.
50. Hahn H, Wojnowski L, Zimmer AM et al. Rhabdomyosarcomas and radiation hypersensitivity in a mouse model of Gorlin syndrome. Nat Med 1998; 4(5):619-622.
51. Goodrich LV, Milenkovic L, Higgins KM et al. Altered neural cell fates and medulloblastoma in mouse patched mutants. Science 1997; 277(5329):1109-1113.

Chapter 5

Stochastic Gene Expression:

Dominance, Thresholds and Boundaries

H. Frederik Nijhout

Abstract

Butterfly color patterns are built as finely tiles mosaics of colored scales, each the product of a single epidermal cell. The overall pattern is composed of a small number of pigments. Each scale appears to make an all-or-none "choice" to synthesize only one of this small set of alternative pigments. Gradual transitions from one color to another on the wing are accomplished by graded changes in the proportions of discretely colored scales, and areas of intermediate coloration are made up of a random salt-and-pepper mosaic of discretely colored scales. In transition regions, and at boundaries between colors, scales also exhibit stochastic patterns of pigment expression. A model for a stochastic mechanism for gene expression is presented that can account for the observed stochastic patterns of pigmentation. This model shows that stochastic gene expression provides a simple mechanism for establishing a step-like, or threshold, response to a spatially graded signal. Conditions that affect the steepness of the threshold are examined. It is shown that stochastic gene expression also results in the emergence of interesting patterns of dominance among the alleles of a gene.

Introduction

Stochastic gene expression occurs when transcriptional regulators are present at very low concentrations, so that binding and release of regulators from their binding sites becomes probabilistic. The suspicion that stochastic gene expression has a significant effect on the biology of organisms comes from the observation that genetically identical (clonal) organisms, maintained in identical environments, diverge phenotypically. For instance, cell division in bacteria growing in an optimum medium rapidly becomes asynchronous, presumably due to individual stochastic variation in regulatory processes.[1] Stochastic gene expression has been observed in artificial genetic constructs.[2] In addition, stochastic gene expression and has been suggested to be a significant cause of haploinsufficiency.[3]

We have found that stochastic gene expression is a common and easily observable feature in the development of butterfly color patterns. Color patterns are made up of pigmented scales; each scale is the outgrowth of a single epidermal cell, and typically contains only a single pigment. The overall color pattern is a pixellated mosaic of monochrome scales, and in most patterns each scale-cell makes an all-or-none decision to synthesize one of a small set of alternative pigments.

Stochastic genes expression occurs in four different ways in this mosaic. First, where alternative colors are controlled by allelic differences at a single gene and there is incomplete dominance, the color of the heterozygote is intermediate between that of the two homozygotes, and

The Biology of Genetic Dominance, edited by Reiner A. Veitia. ©2006 Eurekah.com.

Figure 1. A cross of two genotypes of *Papilio dardanus* with codominant alleles. In many regions of the wing the heterozygote on the right has a pigmentation (gray) roughly intermediate between that of the parental types (black and white). The intermediate pigmentation comes about by a random arrangement of scales of the two parental types (female patterns are shown in this diagram).

this is due, not to an intermediate amount of pigment, but due to the fact that a fraction of the scales develop one color and the remainder develop the other color (Fig. 1). Second, many natural color patterns are entirely made up of a random, salt-and-pepper arrangement of scales of two or three colors, often with gradations in color that are accomplished by spatial changes in the relative frequency of each of the colors (Fig. 2). Thirdly, stochastic gene expression occurs at the boundaries between two areas of solid color. Even the sharpest boundaries are noisy, with many scales expressing a color that is out of place. In many cases the boundaries are rather wide and exhibit a gradient in the frequency of the two adjoining colors (Fig. 3).

Finally, each pigment in the eyespot patterns of butterflies is preceded by the expression of a unique combination of transcription factors.[4] In large areas of homogeneous color, the relevant transcription factors become expressed in all cells, but the initiation of expression is stochastic. Expression begins in a sparse random pattern of cells that gradually fills in until all cells in the field that will become that particular color express these genes.[4] Thus the initial spatial pattern of gene activation in large areas of homogeneous gene expression is also stochastic.

The present paper is an exploration of the potential causes behind these various manifestations of stochastic gene expression. We take as our starting point a mathematical approach to understanding stochastic gene expression and use this to examine the conditions under which the observed patterns of stochastic expression in butterfly wings could occur. We will see that dominance is an emergent property of stochastic gene expression, due to the extreme nonlinearity of the relationship between transactivator concentration and the probability of gene expression.

A General Model of Stochastic Gene Expression

Stochastic gene expression is thought to be the consequence of the random activation and inactivation of transcription due to successive cycles of binding and release of a transcription factor.[5] Activation and inactivation are assumed to occur by a first-order reversible reaction

$$G_{inact} \underset{k_{inact}}{\overset{k_{act}}{\rightleftharpoons}} G_{act}$$

where G represent the gene and where the reaction constants k_{act} and k_{inact} determine the rates of the activation and inactivation reactions, respectively.[3,5,6] Because a gene can be either active or inactive, the rate constants also define the probability that a gene will be activated if inactive, and vice versa.

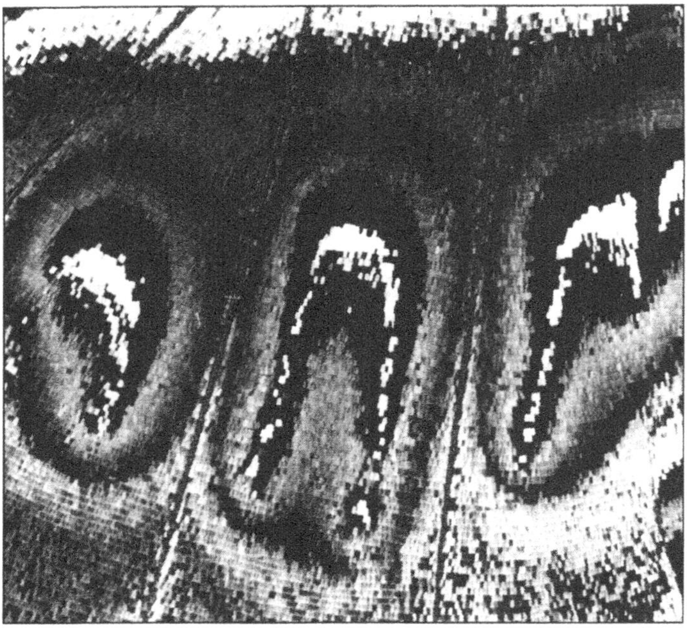

Figure 2. Closeup of the color pattern on the ventral hind wing of *Morpho hecuba*, illustrating stochastic pigmentation at boundaries between pattern and in certain regions of the color pattern.

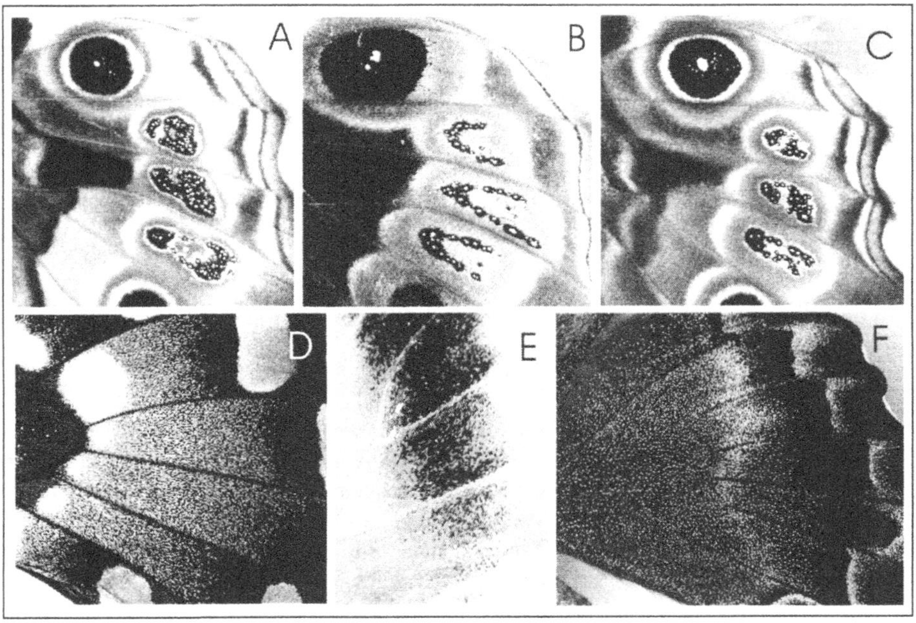

Figure 3. Stochastic pigment expression is a common feature of butterfly color patterns. Top row shows individual variation of stochastic patterning in *Lethe europa*. Other panels illustrate various manifestations of stochastic pigment expression that is part of the normal color pattern of many species of butterflies.

If the probability of gene activation is uniform overt time, then the time between activation events will have an exponential distribution given by the probability density function

$$f(t) = e^{-t/b}/b, \quad (1)$$

and the probability that an event has occurred is given by cumulative density function

$$F(t)= 1- e^{-t/b}, \quad (2)$$

where b is the mean time between events. Among the properties of an exponential distribution is that the standard deviation is the same as the mean, and the median time between events is the half-time of the reaction: $t_{1/2} = \ln2*b = 0.693*b$. If gene activation occurs as a first-order reaction

$$G_{inact} \xrightarrow{k_{act}} G_{act}, \quad \text{so that} \quad dG_{act}/dt = k_{act} * G_{inact} \quad (3)$$

then $k_{act} = 1/b$. The rate constant for the activation reaction thus defines the half-time of gene activation. For example, if $k_{act} = 10^{-4}$ sec^{-1}, then $t_{1/2}$ = 6930 seconds: about 2 hours. The half time of activation is the average time required for half of the genes in a sample to be come activated, or, in the case of single cells, the average time required for one of the two alleles of a gene to become activated. For all simulations reported here time is in arbitrary units (t), rate constants (ks) are in units of t^{-1}. The value of G can be either 0 (G_{inact}) or 1 (G_{act}), and the probability of the $0 \rightarrow 1$ switch is $k_{act} = 1/b = k_1$. The effective value of the rate constant is proportional to ploidy. Thus a diploid has two copies of the gene and thus is able to produce twice as much product in a given period of time. So diploidy can be modeled by doubling the value of k_{act} (assuming both alleles respond identically to the activator) and the consequences for the exponential distribution can be deduced from equation (1).

A gene becomes inactivated by (1) the removal of a transcriptional activator, or by (2) the binding of a transcriptional inhibitor. Like gene activation, these two kinds of inactivation events can also be stochastic. The effects of the two types of inactivation events are presumably independent of each other, so we assume that k_{inact} is the sum of the stochastic rate of removal of the transcriptional activator and the stochastic rate of binding of a transcriptional inhibitor ($k_{inact} = k_{act\text{-}off} + k_{inh\text{-}on}$). The fraction of time a gene is active is given by $k_{act}/(k_{act}+k_{inact})$. This ratio also gives the fraction of cells in which a gene is active at any given time.

We follow ref. 3 in assuming that the activated gene produces a gene product (P) via a first-order mechanism, and that this product decays at a rate proportional to its concentration:

$$dP/dt = k_3*G_{act} - k_4*P. \quad (4)$$

In diploid individuals two genes (G_1 and G_2) contribute independently to the synthesis of product, so the rate becomes

$$dP/dt = k_3*(G_{1act} + G_{2act}) - k_4*P. \quad (5)$$

The concentration of the product P fluctuates with stochastic variation in G_{act} (Fig. 4A). The amplitude of the fluctuation in [P] is determined by the half-times of activation and inactivation of the gene and the rates of synthesis and destruction of P. If the half-time of activation/inactivation is long (implying small k_{act} and k_{inact}) and the rates of P synthesis and decay are large, the fluctuations will be large. If the half-times of inactivation and inactivation are short (implying large k_{act} and k_{inact}) the fluctuation of P will have a smaller amplitude (Fig. 4B).

Nearly steady concentrations of P can be achieved either with very rapid alternation between activation and inactivation, or with extremely slow alternation (so that genes are either on or off most of the time). The mean concentration is determined by all 4 parameters as follows:

$$[P]_{mean} = (G_{1act}+G_{2act})*k_{act}*k_3/(k_{inact}*k_4+k_{act}*k_4), \quad (6)$$

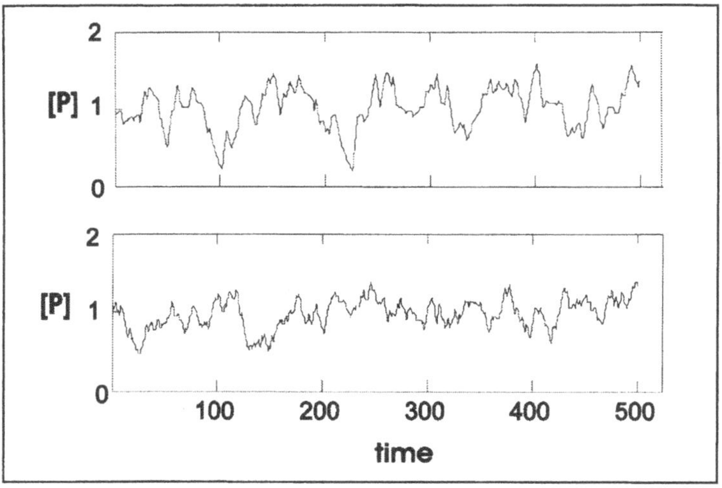

Figure 4. The rate constants of gene activation and inactivation affect the amplitude of the fluctuations in the concentration of gene product. Top panel, activation and inhibition rate constants 0.05/t; bottom panel, activation and inhibition rate constants 0.075/t. X axis is in units of t (time), y axis is concentration of product, P.

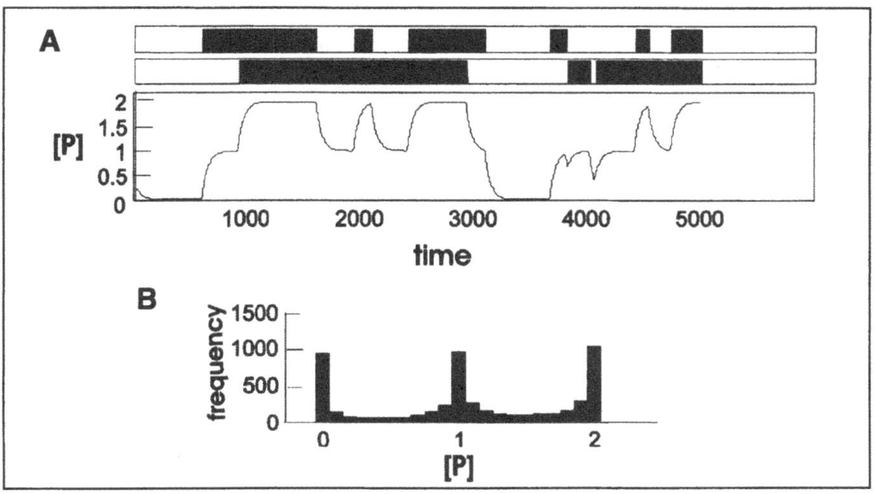

Figure 5. Very low rates of gene activation and inactivation result in product concentrations that dwell at zero, medium and high levels. A) top panel shows bars illustrating the on/off times of two alleles of a diploid above the resulting profile of gene product concentration. B) frequency distribution of gene product concentrations at a standard time point.

where $G_{1act}+G_{2act} = 2$, and k_{act} and k_{inact} are the probabilities that a gene will be activated or inactivated, respectively, in a given time interval. If the values both k_{act} and k_{inact} decrease proportionally, or if the values of both k_3 and k_4 increase proportionally, the mean [P] is unchanged but the system spends an increasing fraction of time at $[P]_{min}$, or $[P]_{max}$, or $1/2[P]_{max}$ (Fig. 5A). The transition between high, medium, and low [P] is random and the probability that the system will be in one of these three states is distributed as shown in Figure 5B.

Stabilizing Gene Expression

We are concerned with two patterns of stochastic gene expression. The first one is in the stochastic initiation of gene expression in a field where gene expression will eventually become homogeneous in all cells. The second is stochastic gene activation that leads to a salt-and-pepper pattern of two alternative stable states of gene expression in different cells (Fig. 2). In both cases it is necessary to turn a randomly fluctuating level of P into a stable state of gene expression that is spatially either homogeneous or heterogeneous. In the case of color patterns, the level of P must unambiguously specify one or another pigment synthesis pathway, so there must be a time point in development when the state of a gene becomes fixed.

One way of stabilizing initially stochastic expression is if the state of gene activation is "read" at a specific point in time, so that the state at that time (active or inactive) becomes fixed. A simple and possibly common way to do this is if [P] activates a process that makes further fluctuation in [P] irrelevant. P could, for instance, be a transcription factor that activates another gene with a more stable activation kinetics. Such a mechanism would require that the response to P behave as a threshold, so that P has no effect below the threshold concentration, and a full activating effect above the threshold. The value of the threshold would then determine the fraction of cells that express the phenotype specified by high levels of P. For instance, if the threshold is below the range of fluctuation of [P], all cells will express the phenotype, and if the threshold is above the range of fluctuation of [P], none will. A threshold within the range of variation in [P] will results in a random expression of alternative phenotypes in different cells. In this scenario, the mechanism that controls the threshold effectively controls the degree of stochasticity of the final pattern.

It is possible, however, to stabilize the expression level of P without the need to assume a threshold mechanism. One way to stabilize the activation of a gene is to reduce k_{inact} to zero (or to a very small value relative to k_{act}) after a period of time. This could be achieved if the product P acts as an inhibitor of the inactivation reaction:

$$dG_{act}/dt = k_{act}^{*} G_{inact} - k_{inact}^{*} G_{act}/P, \qquad (7)$$

so that as [P] rises, inactivation becomes progressively less probable. The mean [P] will approach $[P]_{max}$, but [P] will continue to fluctuate with a low amplitude (Fig. 6).

Increasing the activation reaction relative to the inactivation reaction by means of a positive feedback from P

$$dG_{act}/dt = P^{*} k_{act}^{*} G_{inact} - k_{inact}^{*} G_{act}, \qquad (8)$$

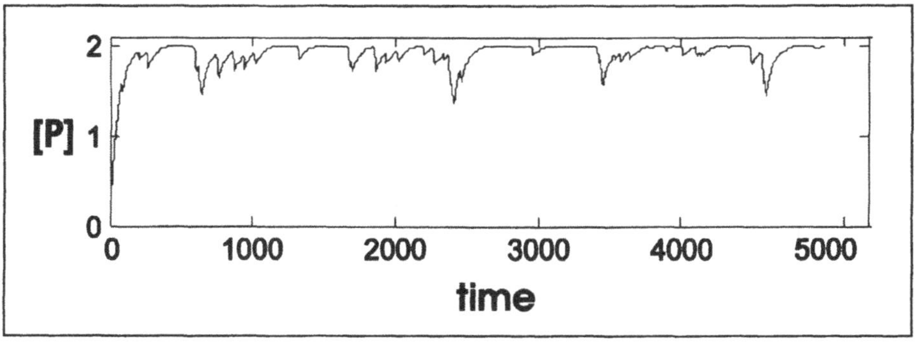

Figure 6. When the stochastic inactivation rate of a gene is much smaller than its stochastic activation rate, the gene product is stabilized and fluctuates with small amplitude at a high level. Here stabilization is achieved through inhibition of gene inactivation by the gene product.

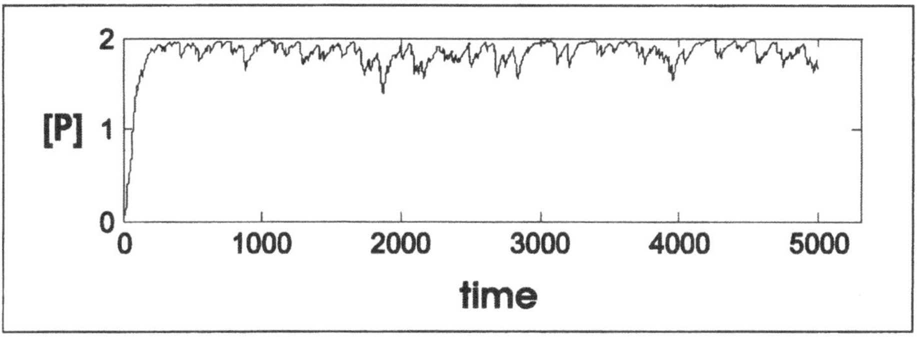

Figure 7. Stabilization of gene product level by positive feedback of gene product to gene activation rate.

has a similar same effect (Fig. 7). Regulation of gene expression by the gene product (autocatalysis) is a common feature of the regulation of genetic regulatory mechanisms. This kind of feedback has one additional feature, in that it makes it possible to permanently inactivate expression of the gene.[3,7] Because gene activity is stochastic, the concentration of the gene product will fluctuate. If the feedback is not too strong then the concentration of the product can drop to zero, and this will effectively inactivate the gene because it brings the activation rate to zero (Equation 8 and Fig. 8). Of course, if the gene is initially inactive, this mechanism prevents it from being activated. Therefore a different stimulus that is not part of the feedback mechanism of P is required to activate gene expression. By itself, this mechanism only stabilizes the inactive state, not the active state, and [P] continues to fluctuate as long as the gene is not inactivated. The time required to become inactivated will have an exponential distribution.

The active state can be stabilized by feedback of P on the inactivation reaction, as suggested above, so a pseudo-bistable system (with [P] either zero or fluctuating at a high level) is

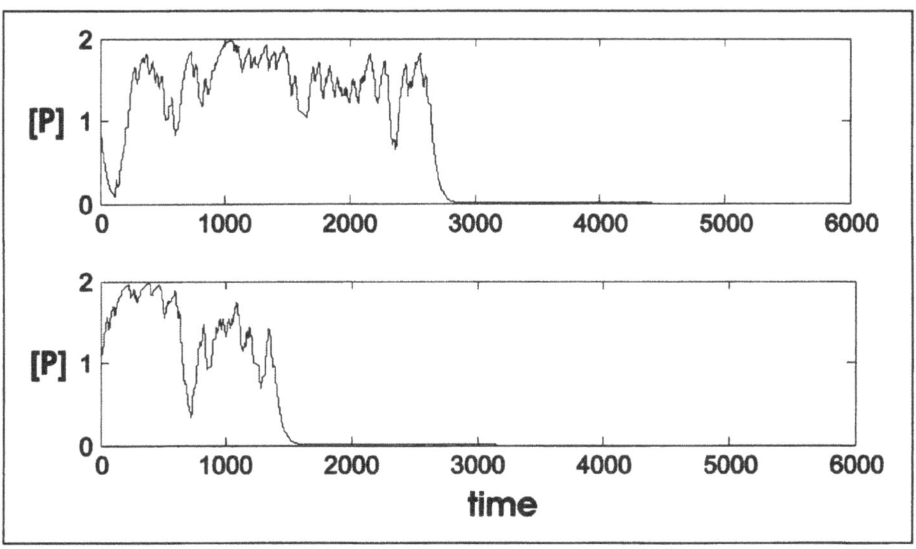

Figure 8. If positive feedback of gene product to gene activation rate is not strong fluctuations are large and can result in permanent inactivation of the gene.

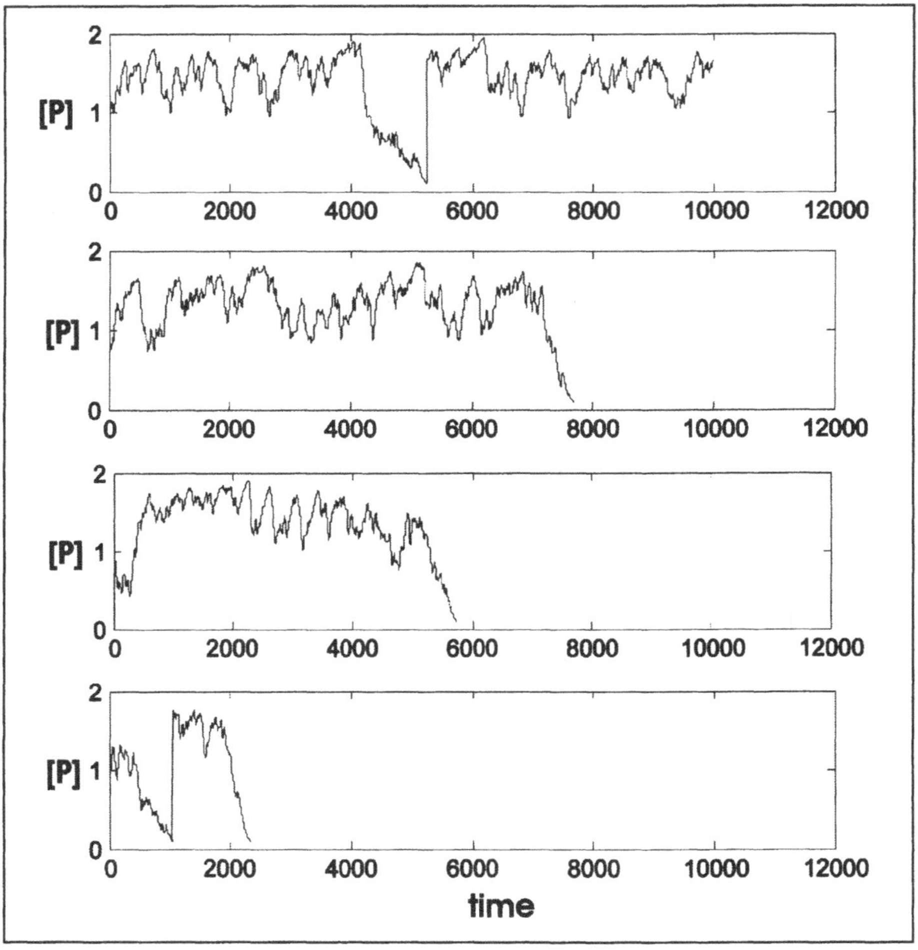

Figure 9. Double feedback results in a pseudo-bistable system in which gene product concentrations fluctuate with relatively low amplitude at a high level with occasional large excursions. When such a large excursion reduces the product concentration to zero the gene is permanently inactivated. The panels show four simulations with identical parameter values.

obtained by double feedback by P on both the activation and inactivation reactions giving (by combining equations 5, 7 and 8):

$$dG_{1act}/dt = P^* k_{act}^* G_{1inact} - k_{inact}^* G_{1act}/P$$

$$dG_{2act}/dt = P^* k_{act}^* G_{2inact} - k_{inact}^* G_{2act}/P \qquad (9)$$

$$dP/dt = k_3^*(G_{1act} + G_{2act}) - k_4^* P$$

Figure 9 illustrates several sample traces of simulations with this set of equations for a diploid system. If an individual is heterozygous for G, so that G_1 is active and G_2 is not, maximum value of P is halved, and the probability that P will decline to zero in a given period of time greatly increased. An example of the distribution of times required for [P] to drop to zero in the

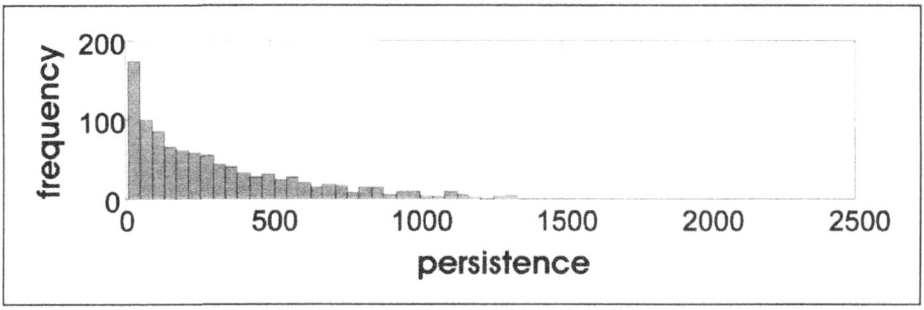

Figure 10. Stochastic gene activation can result in haploinsufficiency. Frequency distribution of 1000 simulations of a haploid system with double feedback illustrating the times at which gene product drops to zero (cf Fig. 9), using the following parameter values: $k_{act} = k_{inact} = 0.1$, $k_3 = k_4 = 0.2$. All genes are inactivated before t = 2,500. In a 1000 simulations of a diploid system with same parameter values the gene product always remained high until t = 10,000.

diploid and haploid condition is shown in Figure 10. The means of these distributions, the percentage of cells with inactivated genes at a given time, and the exact difference between diploids and haploids, depends on the values of the parameters.

Thresholds and Boundaries

During pattern formation in development, medium- and long-distance regulatory signals are often distributed in a smoothly graded fashion, and these graded signals eventually become translated into abrupt localized differences in gene expression. Mechanisms by which a graded signal can be translated into a discrete response have been extensively explored from a theoretical perspective. Most of the mechanisms studied so far establish thresholds or boundaries through some combination of cooperativity, lateral inhibition, bifurcation, and/or diffusive instability.[8,9,10,11,12,13,14]

Cook et al (1998; ref. 3) studied the properties of a stochastic mechanism of gene activation in which an activator stimulated gene transcription with Michaelis-Menten kinetics. They assumed that the accumulation of the product above an arbitrary threshold level stimulated an all-or-none activation of a "phenotype." They showed that this system exhibited a threshold-like response to graded input of the activator, with Hill coefficients ranging from 2.4 to 7.0. They showed that the steepness of the switch depended on the stability of gene activation: faster cycles of gene activation and inactivation produced sharper thresholds than relatively slower rates of gene activation and inactivation. These investigators were primarily interested in examining how the predictability of gene activation affected the response to a graded signal, and suggested that rapid cycling improved the response because it allowed a cell to sample stimulus intensity repeatedly before committing to a particular genetic response.

Here we are interested in the temporal and spatial distribution of the stochastic response. Instead of assuming an arbitrary threshold for the phenotype, we modeled a mechanism that stochastically inactivates the gene permanently or keeps it active with high probability. At a given time in development, the gene product is present or absent with a given probability that depends on the level of transcriptional activators and on the rates of product synthesis and decay.

Spatial variation in parameter values will lead to spatial variation in the probability that the genes in a given cell will be activated or inactivated. Many transcription factors that regulate animal development are distributed as gradients. Smooth gradients of transcription factors are generally translated into more or less discrete spatial differences in the expression of

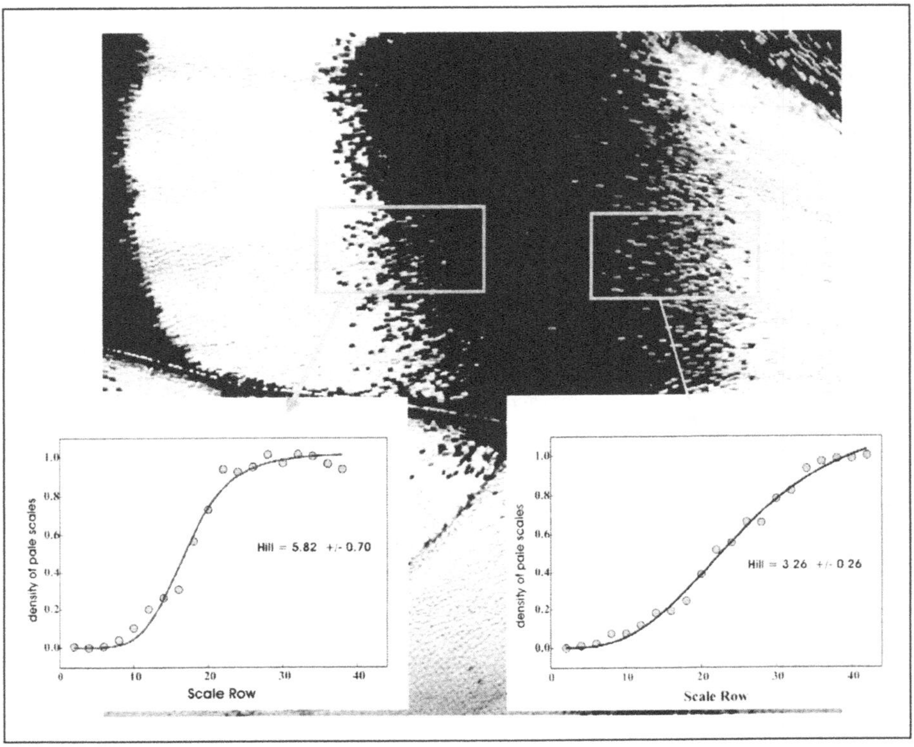

Figure 11. Boundaries between two color fields on the fore wing of *Papilio glaucus*, illustrating stochastic transitions. Colored scales in *P. glaucus*, as in most other butterflies, occur in parallel rows. Graphs show the frequency of light colored scales in each row in transects across two color boundaries. These transitions are sigmoid with rather large Hill coefficients (Hill coefficient calculated by least squares nonlinear regression).

regulated genes. In butterfly color patterns, transition between two colors is never abrupt but resembles a sigmoid transition (e.g., Fig. 11).

In our model, the activity of a transcription factor corresponds to the value k_{act}, the rate constant for gene activation. A linear gradient of a transcription factor is simulated by a linear increase in the value of k_{act}. The effect of such a gradient on stochastic gene activation in the absence of feedback regulation is shown in Figure 12A. The transition from zero to one hundred percent gene activation is sigmoid, with a large Hill coefficient that depends on the choice of parameter values. The response to a linear gradient, in the presence of feedback, is illustrated in Figure 13A. The transition is again sigmoid, with an even larger Hill coefficient. A linear gradient in the distribution of the activity of the enzyme that produces P also yields a sigmoid response in the distribution of gene product expression (Figs. 12B, 13B). The midpoint and steepness of the transition are determined by the shape of the gradient (linear or nonlinear) and the values of the parameters of the gene activation and expression mechanism (Fig. 14).

It is clear from these results that stochastic gene expression, with our without feedback regulation, produces threshold-like responses to graded input signals (see also ref. 3). Stochastic gene regulation is thus inherently a threshold mechanism. In butterfly wing patterns this is revealed by stochastic gradations at boundaries between regions of different colors (Fig. 11). Whether such stochastic transitions are a common feature of boundaries in other biological systems is not known. Experiments in which gene expression is visualized by in situ hybridization

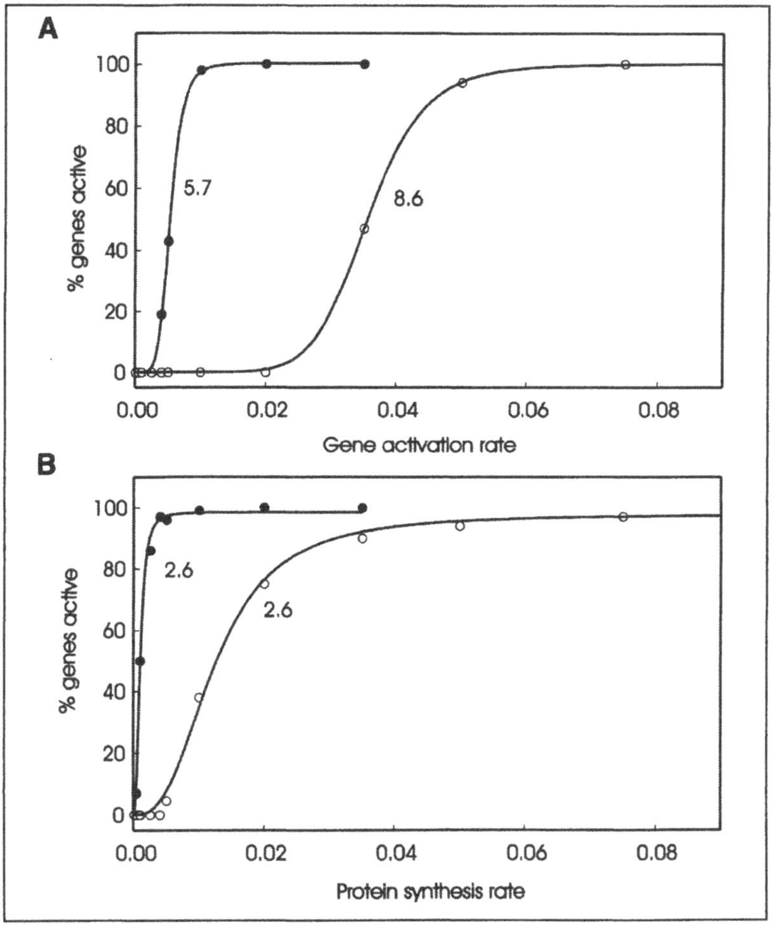

Figure 12. Response of the stochastic gene activation system described by equations (3) and (5) to a linear gradient of transcriptional activator activity (A), or a linear gradient in the activity of a product-synthesizing enzyme (B). Filled circles, all other parameters at 0.01; open circles, all other parameters at 0.05. Ordinate gives the parameter values for the activators or enzyme activity, abscissa gives the percentage of simulations in which the genes were active at t = 10,000. Numbers next to each curve are the Hill coefficients (calculated by nonlinear regression).

almost always reveal substantial stochastic expression at boundaries, so the involvement of stochastic processes in the regulation of thresholds and boundaries may be widespread.

Spatially Homogeneous Stochasticity

Butterflies are evidently able to regulate the stochastic pattern of gene expression in their color pattern with considerable accuracy. In sparse stochastic patterns, like those shown in Figure 3, the density gradients of colored scales are consistent from individual to individual, so these species are evidently able to control a precise proportionality of scales that are either on or off over large regions of their wing. It seems unlikely that these are cases in which the system is poised precisely on a steep transition with a high Hill coefficient, because such a situation would be extremely sensitive to individual variation. It is more likely that these kinds of

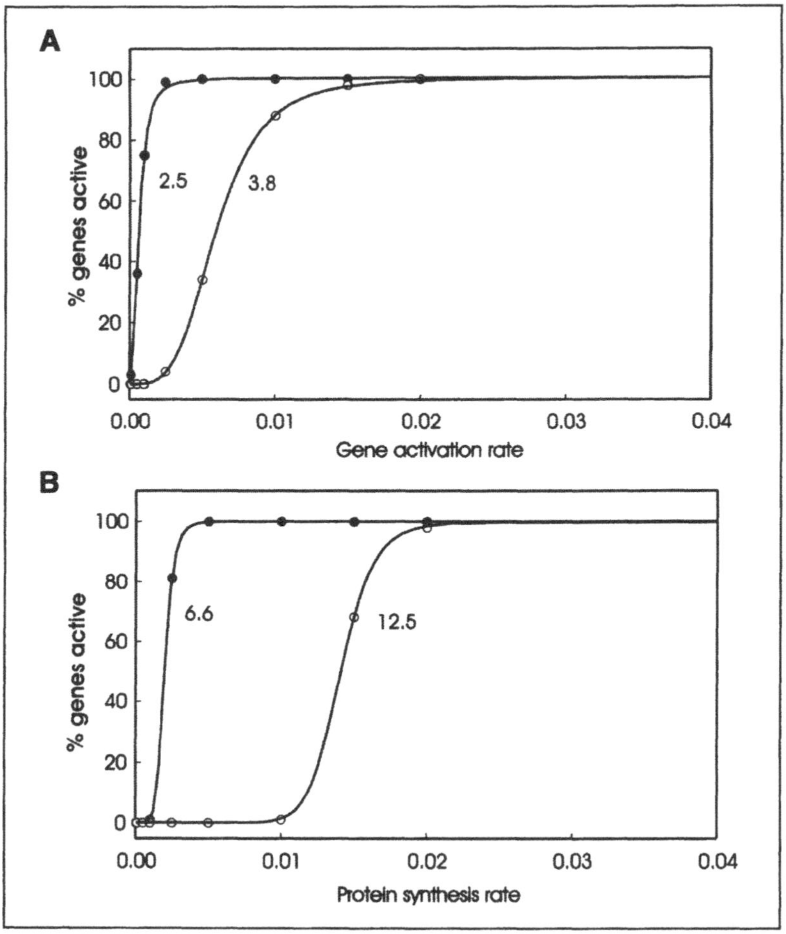

Figure 13. Response of the stochastic gene activation system with double feedback regulation described by equation (9) to a linear gradient of transcriptional activator activity (A), or a linear gradient in the activity of a product-synthesizing enzyme (B). Filled circles, all other parameters at 0.01; open circles, all other parameters at 0.05. Ordinate gives the parameter values for the activators or enzyme activity, abscissa gives the percentage of simulations in which the genes were active at t = 10,000. Numbers next to each curve are the Hill coefficients.

individually invariant stable stochastic patterns exist in a region of geno-space where the Hill coefficients are rather small, such as the region in the foreground of Figure 14.

Thus in butterfly color patterns we see the entire range of stochastic expression, from alternative stability, to more or less sharp transition zones, to broadly distributed spatially homogeneous stochasticity. It is unlikely that each of these instances comes bout through a totally different mechanism. Rather, it is more parsimonious to assume these are all variants of a common mechanism and that natural selection can produce genotypes that occur almost anywhere in the geno-space shown in Figure 14. An interesting possibility raised by this observation on butterflies is that tissues that are homogeneous for one gene activity may be spatially stochastic for another. Whether apparently homogeneous tissues are in fact mosaics of stochastic gene expression is an open question.

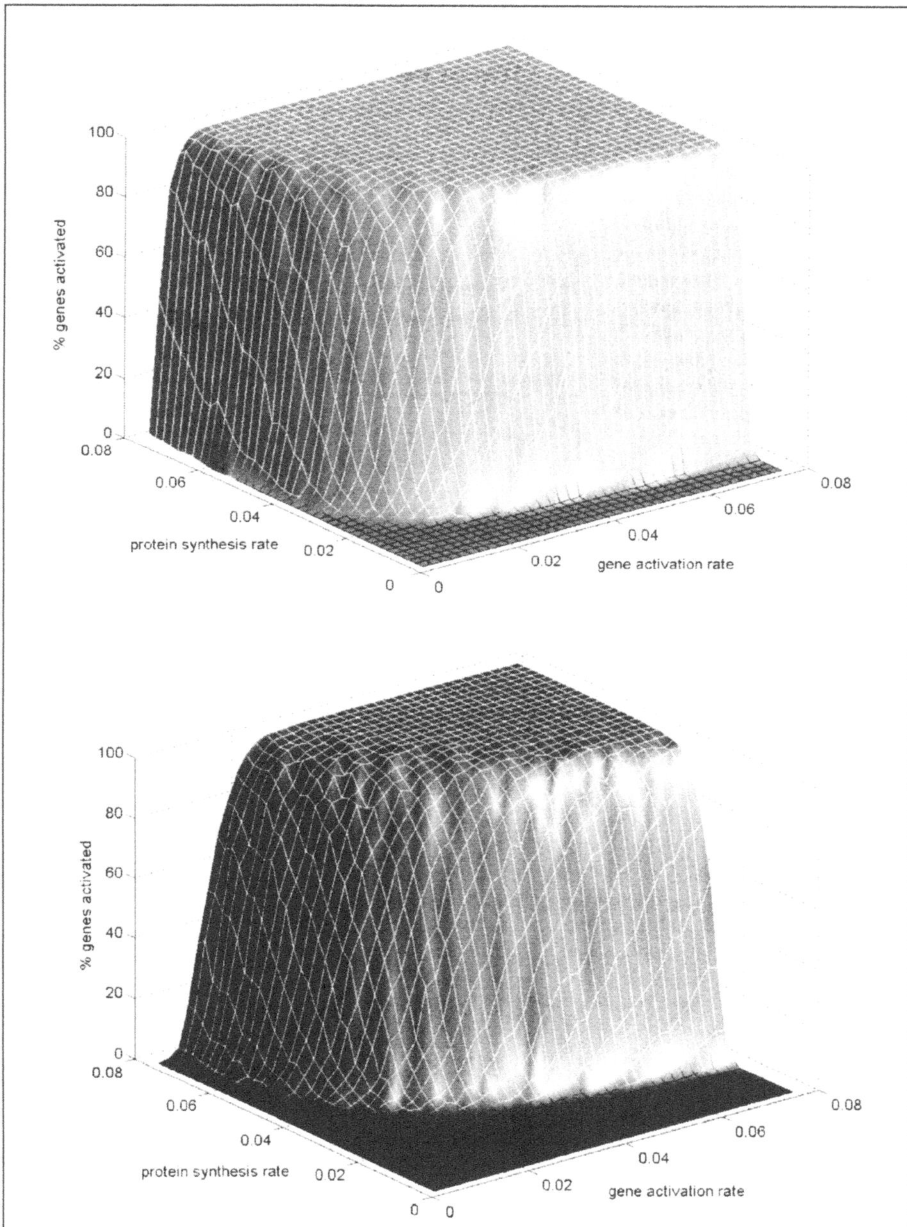

Figure 14. Bivariate plots of the effect of variation in transcription factor activity and product synthesis activity on the level of P. Abscissa shows percentage of cells in which the genes were active at t = 10,000. In the upper flat regions $[P]=[P]_{max}$. In the upper figure all other parameters were set at 0.050, and in the lower figure at 0.075. From these figure it can be seen that the transition are always sigmoid, and that for some combinations of parameter values the transitions are sharp and switch-like, whereas for other combinations the transitions are more gradual (see also Figs. 12, 13). Parameter values also determine the midpoint of the sigmoid transition, and hence the value of the threshold.

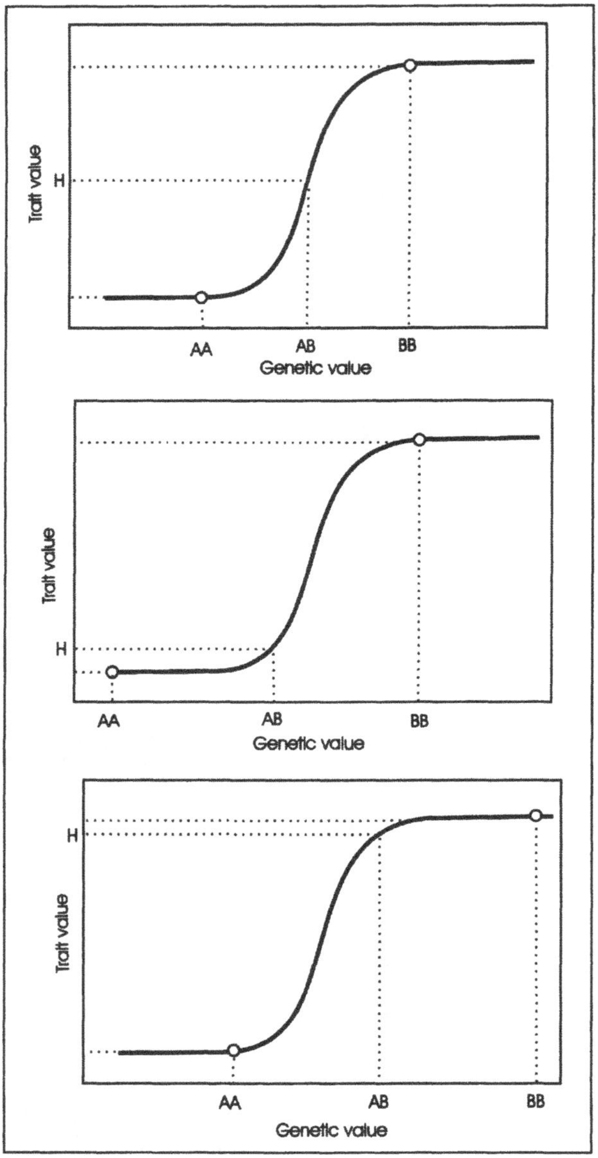

Figure 15. A sigmoid relationship between genotype and phenotype has interesting consequences for the dominance relationships among alleles with different genetic values. In all three panels, open circles indicate the values of two homozygous genotypes. The alleles are assumed to act additively at the physiological level, so the genetic value of the heterozygote (AB) is exactly halfway between that of the two homozygotes (AA and BB). The phenotypic values of the three genotypes are shown with horizontal dotted lines, and the phenotypic value of the heterozygote is indicated by H on the ordinate. Dominance occurs if the phenotypic value of the heterozygote is not halfway between the phenotypic values of the homozygotes. Because of nonlinearity of the relationship, both the absolute and the relative values of the alleles affect their dominance relationships. A hypermorphic allele (on the high plateau) can be either codominant (A), recessive (B), or dominant (C) to a hypomorphic allele.

Dominance

Dominance is an emergent property of the mechanism by which a gene affects the value of a trait. In stochastic gene expression there is a sigmoidal relationship between genetic value and phenotypic value (Figs. 13, 14). This nonlinearity leads to an interesting pattern of dominance among alleles. As shown in Figure 15, alleles that are on either extreme of a sigmoid distribution will be dominant over alleles that are on the transition. Alleles that are on the lower and upper flat regions of the distribution will be codominant only if they are perfectly symmetrical relative to the transition. If they are not, then the allele most distant from the transition will be dominant over the one closer to the transition, and this is independent of whether the allele is hypermorphic or hypomorphic.

Haploinsufficiency in the presence of a null-allele occurs if the remaining 'active' allele has a value that is less than twice the value of the transition midpoint. As can be seen in Figure 14, the value of this midpoint depends on the other genes in the system, so haploinsufficiency is a property of the entire genotype.

Evolution of dominance can occur either by mutational changes in the target gene or by changes in the genetic background.[15] Mutational changes can be represented by the generation of new allelic values along the curve that describes the relationship between genotype and phenotype (e.g., Fig. 15). Changes in the genetic background, by contrast, change the *shape* of the genotype-phenotype curved, in effect changing the slope and inflection point of the sigmoid (for instance the shapes of the curves parallel to the x-axis depend on their position along the y axis in Fig. 14; here the gene represented by the y axis provides the genetic background for the gene represented by the x axis, and vice versa).

Alleles that are located the high or low plateaus of Figure 15 are not under selection, because their variation does not contribute to phenotypic variation. It is unlikely, therefore, that selection on alleles of the focal gene will contribute significantly to the evolution of dominance of that gene, particularly in regions where the transition between the low and high plateaus is very sharp. It is thus more likely that evolution of dominance in systems that are governed by stochastic gene expression occurs by evolution of the genetic background.

References

1. McAdams HH, Arkin A. Stochastic mechanisms in gene expression. Proc Natl Acad Sci USA 1997; 94:814-819.
2. Elowitz MB, Levine AJ, Siggia ED et al. Stochastic gene expression in a single cell. Science 2002; 297:1183-1186.
3. Cook DL, Gerber AN, Tapscott SJ. Modeling stochastic gene expression: Implications for haploinsufficiency. Proc Natl Acad Sci USA 1998; 95:15641-15646.
4. Brunetti C, Paddock S, Carroll S. The generation and diversification of butterfly eyespot color patterns. Curr Biol 2001; 11:1578–1585.
5. Ko MSH. Induction mechanisms of a single gene molecule: Stochastic or deterministic? BioEssays 1992; 14:341-346.
6. Ko MSH. A stochastic model for gene induction. J Theoret Biol 1991; 153:181-194.
7. Kepler TB, Elston TC. Stochasticity in transcriptional regulation: Origins, consequences, and mathematical representations. Biophys J 2001; 81:3116-3136.
8. Turing AM. The chemical basis of morphogenesis. Ser B, London: Phil Trans Roy Soc, 1952; 237:37-72.
9. Lewis J, Slack JMW, Wolpert L. Thresholds in development. J Theor Biol 1977; 65:579-590.
10. Meinhardt H. Models of Biological Pattern Formation. London: Academic Press, 1982.
11. Meinhardt H. Models for positional signaling with applications to the dorsoventral patterning of insects and segregation into different cell types. Development 1989; (Suppl):169-180.
12. Meinhardt H. Organizers and axes formation as a self-organizing process. Int J Dev Biol 2001; 45:177-188.
13. Oster GF. Lateral inhibition models of developmental processes. Math Biosci 1988; 90:265-286.
14. Murray JD. Mathematical Biology. New York: Springer-Verlag, 1989.
15. Gilchrist MA, Nijhout HF. Nonlinear developmental processes as sources of dominance. Genetics 2001; 159:423-432.

CHAPTER 6

Mathematical Models of Haploinsufficiency

Indrani Bose and Rajesh Karmakar

Abstract

We describe simple mathematical models of gene expression to explore the possible origins of haploinsufficiency (HI). In a diploid organism, each gene exists in two copies and when one of these is mutated, the amount of proteins synthesized is reduced and may fall below a threshold level for the onset of some desired activity. This can give rise to HI, a manifestation of which is in the form of a disease. We discuss both deterministic and stochastic models of gene expression and suggest possible scenarios for the occurrence of HI in the two cases. In the stochastic case, random fluctuations around the mean protein level give rise to a finite probability that the protein level falls below a threshold. Increased gene copy number and faster gene expression kinetics reduce the variance around the mean protein level. The difference between slow and fast gene expression kinetics, as regards response to a signaling gradient, is further pointed out.

Introduction

Complex multicellular organisms are in general diploids, i.e., each cell in an organism contains two copies of the full set of genes in contrast to haploids in which each cell contains a single copy of the genome. Genes provide the blueprint for the synthesis of proteins which perform essential functions in cells. If one copy of a gene is mutated, there is approximately a 50% reduction in the level of proteins synthesized. In many cases this does not lead to observable changes and normalcy is retained. A common interpretation of haploinsufficiency (HI) is that it occurs when half normal levels of proteins are insufficient for completing particular tasks, leading to specific types of diseases. More generally, HI may occur when the level of proteins synthesized falls below a critical level for the onset of some desired activity.

There is presently an extensive literature on the genetic and biomedical aspects of HI[1-3] but mathematical models exploring the origin of HI and the issues related to it are practically non-existent. It is by now well-accepted that stochastic processes have considerable effect on patterns of gene expression in cells.[4-8] Cook et al[3] have studied the role of stochastic gene expression in HI by constructing a minimal model of gene expression and using numerical techniques to simulate the model (the same minimal model has been studied in a different context in ref. 8). Their major finding is that when one of the two genes in a diploid organism is inactivated due to mutations, there is an increased susceptibility to stochastic initiations and interruptions of gene expression. As a result, the number of proteins produced may transiently fall below the desired level giving rise to HI. Both increased gene copy number and faster gene expression kinetics reduce expression noise, thus enhancing the possibility of a stable outcome.

A large number of diseases are caused by mutations in genes encoding proteins called transcription factors (TF). More than 30 different human maladies have been attributed to TF HI.[1] TFs regulate gene expression by binding at the promoter region of the gene to be expressed. Cooperative interactions among the TFs favour the formation of bound TF complexes (oligomers). The TFs interact at only one site or at multiple sites of the promoter. A simple mathematical model has been proposed to explore HI in systems involving cooperative assembly of TFs.[2] Such multimeric complexes are essential for initiation of gene expression in many eukaryotic systems. The model explores the relationship of fractional oligomerization Y with the free ($[S]$) as well as total ($[S_0]$) concentrations of TFs. The TFs oligomerise to form a bound complex. The curves Y versus $[S]$ and $[S_0]$ have sigmoidal shapes. Due to the characteristic S shape of a sigmoid, a small change in the TF concentration around the inflection point (the point at which the tangent to the curve has the maximum slope) gives rise to a significant change in the magnitude of Y. Thus, if there are two TF-encoding genes and one of these becomes silent, the level of TFs produced may fall below the inflection point of the sigmoid and consequently the magnitude of Y, the fractional oligomerization, is considerably decreased. This results in reduced expression from the target gene, giving rise to TF HI if the amount of proteins synthesized falls below a threshold level.

In the second section of this Chapter, we describe an extended version of the minimal model of Cook et al[3] to investigate the influence of bound complexes of TFs on the initiation of gene expression. The effect is studied in the deterministic rate equation approach. In the third section, we discuss the stochastic version of the minimal model and its extensions to elucidate the role of stochasticity in HI. Stochasticity introduces fluctuations around the mean protein level giving rise to a distribution of protein levels. In the deterministic approach, the effect of fluctuations is ignored and one obtains only the mean protein level. This level may be above the threshold level for the onset of protein activity in which case the possibility of HI does not occur. Inclusion of stochastic effects leads to a broadened protein distribution so that the protein level may transiently fall below the threshold even if the mean level is above it. Such transient excursions can give rise to irreversible effects leading to HI.

Deterministic Model

The model is an extension of the minimal model of gene expression studied by Cook et al.[3] A brief description of the model is as follows. A gene can be in two possible states: inactive (G) and active (G^*). Random transitions occur between the states G and G^* according to the first order reaction kinetics

$$G \underset{k_d}{\overset{k_a}{\rightleftharpoons}} G^* \xrightarrow{j_p} p \xrightarrow{k_p} \Phi \qquad (1)$$

where k_a and k_d are the activation and deactivation rate constants. The corresponding half-times are $T_a = log2/k_a$ and $T_d = log2/k_d$ respectively. In the active state G^*, the gene synthesizes a protein (p) with the rate constant j_p. The protein product degrades with a rate constant k_p and the associated half-time is T_p. The protein degradation product is represented as Φ. The model (Eq. (1)) describes constitutive gene expression. We now assume that activation from the state G to the state G^* is brought about by an inducing stimulus S, e.g., TFs. The reaction scheme in the presence of the stimulus is given by

$$G+S \underset{k_2}{\overset{k_1}{\rightleftharpoons}} GS \underset{k_d}{\overset{k_a}{\rightleftharpoons}} G^* \xrightarrow{j_p} p \xrightarrow{k_p} \Phi \qquad (2)$$

where GS represents the bound complex of G and S from which transition to the active state G^* occurs. If n_G is the total concentration of genes then

$$n_G = [G] + [GS] + [G^*] \tag{3}$$

where $[G]$, $[GS]$ and $[G^*]$ denote the concentrations of genes in the states G, GS and G^* respectively. In the steady state, we have

$$\frac{d[G]}{dt} = k_2[GS] - k_1[G][S] = 0$$

so that

$$\frac{[G][S]}{K_1} = [GS] \tag{4}$$

where $K_1 = k_2/k_1$ is the equilibrium dissociation constant. From (3) and (4), we get

$$[GS] = \frac{n_G[S]/K_1}{1+[S]/K_1} - [G^*]\frac{[S]/K_1}{1+[S]/K_1} \tag{5}$$

Also, in the steady state,

$$\frac{d[G^*]}{dt} = k_a[GS] - k_d[G^*] = 0 \tag{6}$$

From (5) and (6), the expression for $[G^*]$ in the steady state is given by

$$[G^*] = \frac{n_G k_a \frac{[S]/K_1}{1+[S]/K_1}}{k_a \frac{[S]/K_1}{1+[S]/K_1} + k_d} \tag{7}$$

The reaction scheme in (1) leads to the expression

$$[G^*] = \frac{n_G k_a}{k_a + k_d} \tag{8}$$

in the steady state. Expressions (7) and (8) are equivalent on defining effective activation and deactivation rate constants:

$$k_a' = k_a \frac{[S]/K_1}{1+[S]/K_1}$$

$$k_d' = k_d \tag{9}$$

These equivalence relations are useful as one can then map the reaction scheme in (2) onto the simpler scheme in (1) while calculating various quantities. Use of the simpler reaction scheme leads to greater mathematical tractability.

We now assume the inducing stimulus to be TFs. In the simplest approximation, n individual TFs oligomerise to produce an active complex S_n according to the reaction scheme

$$nS \underset{}{\overset{K}{\rightleftharpoons}} S_n \tag{10}$$

The n TFs interact all at once to give rise to the bound complex $[S_n]$, i.e., we ignore the formation of dimers, tetramers, ... etc. Let $[S_0]$ be the initial concentration of TFs. Then

$$[S_0] = [S] + n[S_n] \tag{11}$$

where $[S]$ and $[S_n]$ are the concentrations of free TFs and the bound TF-complex respectively. The global equilibrium constant K is given by

$$K = \frac{[S_n]}{[S]^n} = \frac{[S_n]}{\left([S_0] - n[S_n]\right)^n} \tag{12}$$

The fractional oligomerization is defined as[2]

$$Y = \frac{n[S_n]}{[S_0]} = \frac{n[S_n]}{[S] + n[S_n]} \tag{13}$$

Using (11) and (12), Y can further be written as

$$Y = \frac{[S]^n}{\frac{[S]}{Kn} + [S]^n} \tag{14}$$

and

$$Y = \frac{\left([S_0] - n[S_n]\right)^n}{\frac{([S_0]-n[S_n])}{Kn} + \left([S_0] - n[S_n]\right)^n} \tag{15}$$

Figure 1 shows the curve, fractional oligomerization Y versus $[S_0]$ (Eq. (15)) for $n = 6$, $K = 2$ and $[S_n] = 0.2$. The curve has the well-known sigmoidal shape.

We now replace $[S]$ by $[S_n]$ in the reaction scheme described by Eq. (2), i.e., we assume that the TF-oligomer S_n binds to a gene in the inactive state G to give rise to the bound complex GS_n. Transition to the active state G^* occurs from the intermediate state GS_n. The concentration $[G^*]$ in the steady state is obtained from (7) by replacing $[S]$ by $[S_n]$ where $[S_n] = K[S]^n$ (Eq. (12)). One finally obtains

$$[G^*] = \frac{n_G k_a \frac{K[S]^n / K_1}{1 + K[S]^n / K_1}}{k_a \frac{K[S]^n / K_1}{1 + K[S]^n / K_1} + k_d} \tag{16}$$

Again, (16) reduces to the simpler form in (8) by definig effective rate constants. The concentration of proteins in the steady state is given by

$$[p] = \frac{j_p}{k_p}[G^*] \tag{17}$$

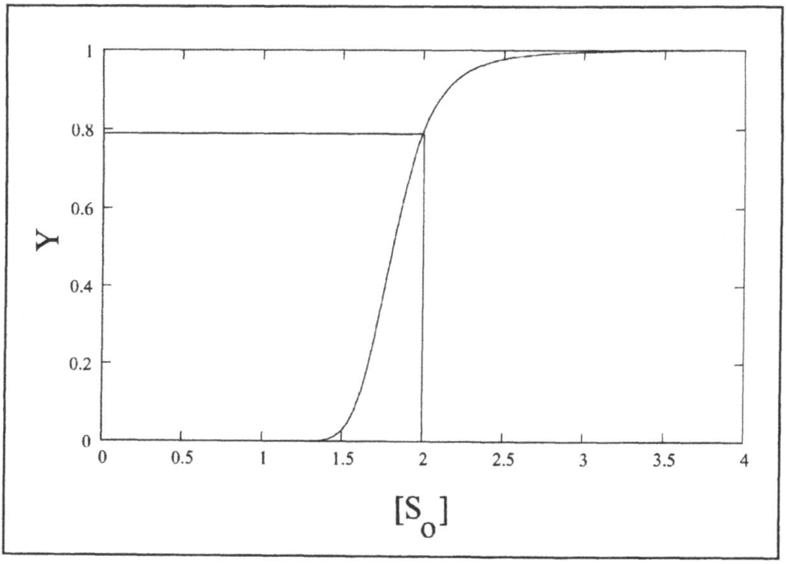

Figure 1. Fractional oligomerization Y versus $[S_0]$ (Eq. (15)) for $n = 6$, $K = 2$ and $[S_n] = 0.2$.

From (14), $[S]^n$ can be written as

$$[S]^n = \frac{Y}{1-Y}\,\frac{[S]}{Kn} \tag{18}$$

From (16) and (17) and for $k_a = k_d$, we get

$$[p] = n_G \frac{j_p}{k_p}\,\frac{aY}{1+2aY-Y} \tag{19}$$

where $a = [S]/(n\,K_1)$.

Figure 2 shows the protein concentration $[p]$ versus fractional oligomerization Y for $k_a = k_d$ (Eq.(19)), $n_G = 2$, $j_p = 0.5\,k_p$, $K_1 = 1.2$ and $n = 6$. From Eqs. (11), (16), and (17) and for $k_a = k_d$, the concentration of protein $[p]$, as a function of the total concentration $[S_0]$ of TFs, can be written as

$$[p] = n_G \frac{j_p}{k_p}\,\frac{\left([S_0]-n[S_n]\right)^n}{2\left([S_0]-n[S_n]\right)^n + K_1/K} \tag{20}$$

Figure 3 shows the plot $[p]$ versus $[S_0]$ (Eq. (20)) for $n = 6$, $K = 2$, $K_1 = 1.2$ and $[S_n] = 0.2$. Figures 1, 2 and 3 provide a possible explanation for the origin of HI. Suppose two TF-encoding genes produce TFs of total concentration $[S_0] = 4$. If one of the genes is inactivated due to mutations, the total concentration $[S_0]$ falls to the value 2. For the parameter values corresponding to Figure 1, the fractional oligomerization Y has the value 0.797. The TFs form a bound complex S_n ($n = 6$) which then activates the gene synthesizing the protein p. The concentration of proteins $[p]$ corresponding to $Y = 0.797$ and for $[S] = 0.8$ is given by $[p] = 0.233$ (Fig. 2). The same value $[p]$ is obtained from Figure 3 with $[S_0] = 2$. If both the TF encoding genes are active, the values of $[S_0]$, Y and p are $[S_0] = 4$, $Y = 1.0$ and $[p] = 0.5$ respectively. Thus, for one gene, the protein level is reduced by more than half. As pointed out by Veitia,[2] marked

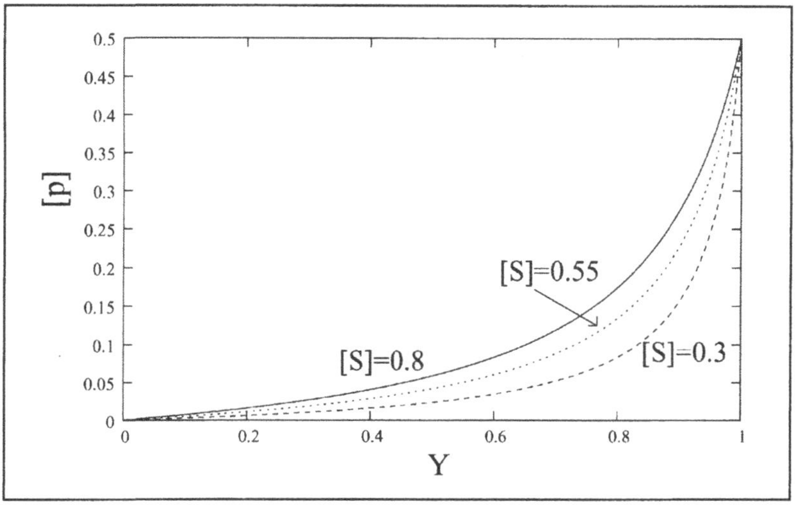

Figure 2. Protein concentration $[p]$ versus fractional oligomerization Y for $k_a = k_d$ (Eq. (19)), $n_G = 2$, $j_p = 0.5k_p$, $K_1 = 1.2$ and $n = 6$.

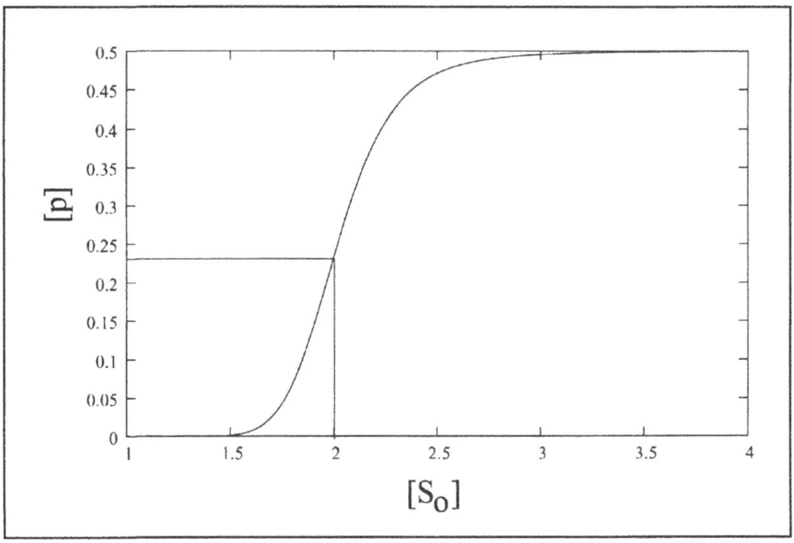

Figure 3. Protein concentration $[p]$ versus $[S_0]$ (Eq. (20)) for $n = 6$, $K = 2$, $K_1 = 1.2$ and $[S_n] = 0.2$.

changes in protein levels are obtained if the stimulus (TF) concentration $[S_0]$ falls in a region around the inflection point of the sigmoid. Suppose the original concentration of $[S_0]$ is $[S_0] = 3.5$ when two genes are synthesizing the TFs. If one of the genes becomes inactive due to mutations, $[S_0]$ falls to a concentration level 1.75. The amount of proteins synthesized from the target genes is then reduced to the considerably lower value of ≈ 0.05 (Fig. 3). If the level of proteins synthesized falls below a threshold, the amount of proteins available is not sufficient for the execution of a particular task. This may give rise to HI, the manifestation of which is in the form of disease. Nonlinearity is responsible for the considerable reduction in protein level when one of the two TF-encoding genes is inactivated. In the minimal model of constitutive gene expression (Eq. (1)), halving n_G (Eq. (8)) leads to a halving of protein concentration $[p]$ (Eq. (17)) in the steady state. Thus, changes in this case are linearly proportional.

To sum up, we have considered a deterministic model in which a complex of n TFs binds at the appropriate region of DNA to initiate gene expression in eukaryotes. The concentration of proteins synthesized versus the total concentration of TFs (Fig. 3) is a sigmoid. Due to the *S*-shape of the curve, the protein level may fall below a threshold when one of the two genes synthesizing the TFs is mutated, resulting in a 50% reduction in the total concentration of TFs. The absence of essential protein activity can give rise to HI.

Stochastic Approach

Gene expression involves several biochemical reactions/events, the kinetics of which determine how the number of participating biomolecules changes as a function of time. In the traditional rate-equation approach, the relevant variables are the concentrations of different biochemical entities or molecules. The time evolution of the concentration variables is assumed to be continuous and deterministic. In reality, the time evolution is not a continuous process as molecular population levels in a reacting system change only by discrete integer amounts. Furthermore, the time evolution is not deterministic as the biochemical reactions/events are inherently probabilistic in nature. This is enhanced by the fact that in diploids two gene copies per cell are actively involved when there is no gene redundancy. For example, the time at which *RNA* polymerase gets bound to the promoter region of a gene or is dissociated from it cannot be predicted with certainty. This gives rise to fluctuations in reaction rates so that the number of *mRNAs* and proteins synthesized by identical genes may vary from gene to gene. The number fluctuations are considerable when a small number of biomolecules participate in the reactions. In the limit of large numbers, fluctuations are much smaller in magnitude so that the deterministic approach is well-justified.

We now consider the model described in the second section, in the absence of an inducing stimulus, and discuss how stochasticity affects gene expression. Let n_{tot} be the total no of genes (gene copy number) and n_0, n_1 ($n_{tot} = n_0 + n_1$), the number of genes in the inactive (G) and active (G^*) states respectively. In the stochastic model, a gene makes random transitions between the inactive and active states with k_a and k_d being the activation and deactivation rate constants. In the active state, protein production and degradation occur with the rate constants j_p and k_p respectively. Let $p(n_1, n_2, t)$ be the probability that at time t, n_1 genes are in the active state G^* and the number of protein molecules is n_2. The rate of change of the probability with respect to time is given by the Master Equation

$$\begin{aligned}\frac{\partial p(n_1,n_2,t)}{\partial t} &= k_a\left[(n_{tot}-n_1+1)\,p(n_1-1,n_2,t)-(n_{tot}-n_1)\,p(n_1,n_2,t)\right]\\&+k_d\left[(n_1+1)\,p(n_1+1,n_2,t)-n_1\,p(n_1,n_2,t)\right]\\&+j_p\left[n_1\,p(n_1,n_2-1,t)-n_1\,p(n_1,n_2,t)\right]\\&+k_p\left[(n_2+1)\,p(n_1,n_2+1,t)-n_2\,p(n_1,n_2,t)\right]\end{aligned} \tag{21}$$

For each rate constant, there is a gain term which adds to the probability and a loss term which subtracts from the probability. The Master Equation is a rate equation in which probability replaces concentration as the relevant variable. We now give a physical interpretation of the probability $p(n_1, n_2, t)$. Consider a collection of identical systems, say, cells with the same initial conditions. Let $P(n_1, n_2, t)$ be the number of cells in which there are exactly n_1 genes in the active state and n_2 number of proteins at time t. This number can increase as well as decrease due to the occurrence of biochemical events. In (21), these events are gene activation, gene deactivation, protein synthesis and protein degradation with rate constants k_a, k_d, j_p and k_p respectively. For the sake of illustration, consider the last term in (21). The number $P(n_1, n_2, t)$ can increase in the infinitesimal time interval $(t, t + \Delta t)$, if in the same time interval a protein decay occurs in a cell containing n_1 genes in the active state and $(n_2 + 1)$ number of proteins at time t. In the infinitesimal time interval Δt, the total number of such cells in which a protein decays is given by $k_p(n_2 + 1)P(n_1, n_2 + 1, t)\Delta t$. Thus if one consider only the gain terms, one gets

$$P(n_1,n_2,t+\Delta t)=P(n_1,n_2,t)+k_p(n_2+1)P(n_1,n_2+1,t)\Delta t$$

In the limit of $\Delta t \to 0$,

$$\frac{\partial P(n_1,n_2,t)}{\partial t}=\lim_{\Delta t\to 0}\frac{P(n_1,n_2,t+\Delta t)-P(n_1,n_2,t)}{\Delta t}$$
$$=k_p(n_2+1)P(n_1,n_2+1,t)$$

The loss term can be obtained in a similar manner. The probability $p(n_1, n_2, t)$ is obtained by dividing the number $P(n_1, n_2 + 1, t)$ by the total number of cells and is ≤ 1.

We now use the standard approach in the theory of stochastic processes[9] to determine the average number of activated genes $< n_1 >$ and proteins $< n_2 >$ in the steady state and the variances thereof. Define the generating function

$$F(z_1,z_2,t)=\sum_{n_1,n_2} z_1^{n_1} z_2^{n_2}\, p(n_1,n_2,t) \tag{22}$$

In terms of the generating function, the Master equation (21) becomes

$$\frac{\partial F}{\partial t}=k_a n_{tot}(z_1-1)F-k_a(z_1-1)z_1\frac{\partial F}{\partial z_1}-k_d(z_1-1)\frac{\partial F}{\partial z_1}+j_p(z_2-1)z_1\frac{\partial F}{\partial z_1}-k_p(z_2-1)\frac{\partial F}{\partial z_2} \tag{23}$$

In the steady state $\partial F/\partial t = 0$. The following properties of the generating function are used in subsequent calculations:

$$F\big|_{z_1=1,z_2=1}=1 \tag{24}$$

$$< n_1 > \;=\; \frac{\partial F}{\partial z_1}\Big|_{z_1=1,z_2=1}, \quad < n_2 > \;=\; \frac{\partial F}{\partial z_2}\Big|_{z_1=1,z_2=1} \tag{25}$$

where $< n_1 >$ is the mean number of active genes, i.e., genes in the state G^* and $< n_2 >$ is the same for proteins. Eq. (24) arises from the fact that the sum of all probabilities should be unity. The average of a quantity, say, protein number is given by

$$< n_2 > \;=\; \sum_{n_1,n_2} n_2\, p(n_1,n_2,t)$$

Different values of n_1 and n_2 define different states of the system with $0 \leq n_1 \leq n_{tot}$ and $0 \leq n_2 \leq n_2^{max}$ where n_2^{max} is the maximum number of proteins synthesized. To obtain the average, which is a weighted sum, one multiplies the protein number n_2 by the probability $p(n_1, n_2, t)$ that the system exists in a state with protein number n_2 and then a summation is taken over all the states of the system.

Furthermore,

$$\begin{aligned} \frac{\partial^2 F}{\partial z_1^2}\Big|_{z_1=1, z_2=1} &= <n_1^2> - <n_1> \\ \frac{\partial^2 F}{\partial z_2^2}\Big|_{z_1=1, z_2=1} &= <n_2^2> - <n_2> \end{aligned} \tag{26}$$

Hence the variances around the mean levels are given by

$$\begin{aligned} Var_{n_1} &= <n_1^2> - <n_1>^2 = \frac{\partial^2 F}{\partial z_1^2}\Big|_{z_1=1, z_2=1} + <n_1> - <n_1>^2 \\ Var_{n_2} &= <n_2^2> - <n_2>^2 = \frac{\partial^2 F}{\partial z_2^2}\Big|_{z_1=1, z_2=1} + <n_2> - <n_2>^2 \end{aligned} \tag{27}$$

Successive differentiation of Eq. (23) ($\partial F / \partial t = 0$) with respect to z_1 and z_2 gives rise to linear equations for successively higher moments. The equations may be solved to obtain, in particular, the mean and the variance. For example, differentiating Eq. (23) with respect to z_1 and z_2 and then putting $z_1, z_2 = 1$ one obtains expressions for the mean.

The mean and variance are given by

$$<n_1> = \frac{n_{tot} k_a}{k_a + k_d} \tag{28}$$

$$Var_{n_1} = <n_1> \frac{k_d}{k_a + k_d} \tag{29}$$

$$<n_2> = <p> = <n_1> \frac{j_p}{k_p} = \frac{j_p}{k_p} \frac{n_{tot} k_a}{k_a + k_d} \tag{30}$$

$$Var_{n_2} = <n_1> \frac{j_p}{k_p}\left[1 + \frac{j_p k_d}{(k_a + k_d)(k_a + k_d + k_p)}\right] \tag{31}$$

As in ref. 3, temporal quantities are scaled relative to the product half-life $T_p = log2/k_p$. Let $T_a = log2/k_a$ and $T_d = log2/k_d$ be the times for half-maximal gene activation and deactivation respectively. The times T_a and T_d are scaled relative to T_p. Some of the results obtained in ref. 3, using numerical simulation techniques, can readily be derived from the analytical expressions in (28)-(31). Stochasticity introduces random fluctuations around the mean protein level and variance gives a measure of the spread. Let $T_a = T_d = T_p/4$, i.e., $k_a = k_d = \alpha k_p$ with $\alpha > 0$. As α increases, one has faster expression kinetics and from (31) it is easy to verify that variance is reduced, i.e., the expression noise is less. The mean product level (Eq. (30)) is, however, independent of α. With increase in j_p, i.e., the protein synthesis rate, the variance increases. Let us now consider the case when the net expression rate of n_{tot} genes is distributed to one single gene so that the mean protein level remains the same. From (30)

$$<n_2> = \frac{j_p}{k_p}\frac{n_{tot}}{2} = \frac{j'_p}{k_p}\frac{1}{2} \tag{32}$$

where $j'_p = j_p n_{tot}$ is the expression rate when only one gene is considered. Since $j'_p > j_p$, the variance is increased on reducing the gene copy number. Similarly, when the net expression rate of n_{tot} genes is distributed to a larger number genes (existence of gene duplicates or paralogs with the same function), say, from two to four, the variance is reduced. When one of two genes is inactivated due to mutations, the average protein level in the steady state is reduced by 50%. This may still be higher than the threshold level required for protein activity. Due to the variance around the mean level, the number of proteins may transiently fall below the threshold giving rise to HI. The occurrence of HI further becomes more probable for slower expression kinetics as then the variance is increased in magnitude. For stochastic gene expression in the presence of an inducing stimulus, say, TF's, we use the effective model with the activation/deactivation rate constants given in Eq. (9). The expressions for the mean and the variance are the same as in Eqs. (28)-(31) but with k_a, k_d replaced by k'_a and k'_d respectively.

We now derive expressions for the probability distributions of protein levels in the steady state. To do this, we consider a simpler stochastic model in which the only stochasticity arises from the random transitions of a gene between the inactive and active states as in the minimal model of Cook et al.[3] In this case, protein synthesis from the active gene and protein degradation occur in a deterministic manner. In the more general model described by Eq.(21), protein production and degradation are treated as stochastic processes. In each state of the gene, the concentration of proteins evolves deterministically according to the equation

$$\frac{dx}{dt} = \frac{j_p}{X_{\max}} z - k_p x = f(x,z) \tag{33}$$

where $z = 1$ (0) when the gene is in the active (inactive) state and $x = X/X_{max}$, X and X_{max} being the protein concentration at time t and the maximum protein concentration respectively. The variable x thus denotes protein concentration normalized by the maximum possible concentration. We note that $X_{max} = j_p/k_p$. Let $p_j(x, t)$ ($j = 0, 1$) be the probability density function when $z = j$. The total probability density function is

$$p(x,t) = p_0(x,t) + p_1(x,t) \tag{34}$$

The rate of change of probability density is given by

$$\frac{\partial p_j(x,t)}{\partial t} = -\frac{\partial}{\partial x}\left[f(x,j)p_j(x,t)\right] + \sum_{k \neq j}\left[W_{kj}p_k(x,t) - W_{jk}p_j(x,t)\right] \tag{35}$$

where W_{kj} is the transition rate from the state k to the state j and W_{jk} is the same for the reverse transition. The first term in Eq. (35) is the so called "transport" term representing the net flow of the probability density. The second term represents the gain/loss in the probability density due to random transitions between the state j and other accessible states. In the present case, Eq. (35) gives rise to the following two equations:

$$\frac{\partial p_0(x,t)}{\partial t} = -\frac{\partial}{\partial x}\left(-k_p x p_0(x,t)\right) + k_d p_1(x,t) - k_a p_0(x,t) \tag{36}$$

$$\frac{\partial p_1(x,t)}{\partial t} = -\frac{\partial}{\partial x}\left\{\left(\frac{j_p}{X_{\max}} - k_p x\right)p_1(x,t)\right\} + k_a p_0(x,t) - k_d p_1(x,t) \tag{37}$$

The steady state distribution ($\partial p_0(x, t)/\partial t = 0$, $\partial p_1(x, t)/\partial t = 0$) in this case is given by

$$p(x) = C x^{\left(\frac{k_a}{k_p}-1\right)}(1-x)^{\left(\frac{k_d}{k_p}-1\right)} \tag{38}$$

where C the normalization constant is given by the inverse of a beta function.[10]

$$C = \frac{1}{B\left(\frac{k_a}{k_p}, \frac{k_d}{k_p}\right)} \tag{39}$$

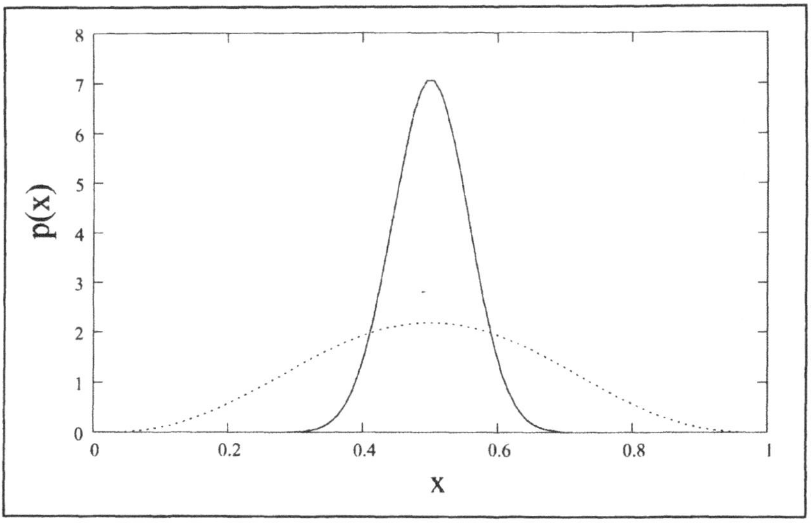

Figure 4. $p(x)$ versus x for slow (dotted curve) and fast (solid) gene expression kinetics.

Since the probability density function is known, the mean protein level and its variance can be calculated in a straightforward manner. The mean value of x, a continuous variable, is given by

$$<x> = \int_0^1 x\,p(x)dx$$

The mean protein level is identical in form to that obtained from the Master equation (Eq. (21)) whereas the variance is underestimated as stochasticity is taken into account only at the levels of gene activation and deactivation. Figure 4 shows the plot of $p(x)$ versus x for slow ($T_a = T_d = T_p/4$) and fast ($T_a = T_d = T_p/40$) gene expression kinetics. In the latter case, the distribution is significantly narrower, i.e., faster kinetics lead to a reduction in the variance. The same conclusion is reached from the Master equation approach. From the full width at half maximum of the broader distribution, one finds that x ranges from 0.25 to 0.75, 0.5 being the mean value. Thus, it is probable that the protein level falls below the threshold for desired activity giving rise to HI. Let x_{thr} (<1) be the threshold value of x. The probability that x is greater than x_{thr} is

$$p(x > x_{thr}) = 1 - \frac{\int_0^{x_{thr}} x^{\left(\frac{k_a}{k_p}-1\right)}(1-x)^{\left(\frac{k_d}{k_p}-1\right)}dx}{\int_0^1 x^{\left(\frac{k_a}{k_p}-1\right)}(1-x)^{\left(\frac{k_d}{k_p}-1\right)}dx} \quad (40)$$

$$= 1 - \frac{k_p x_{thr}^{\frac{k_a}{k_p}}\,{}_2F_1\left[1-\frac{k_d}{k_p}, \frac{k_a}{k_p}, 1+\frac{k_a}{k_p}, X_{thr}\right]}{k_a B\left(\frac{k_a}{k_p}, \frac{k_d}{k_p}\right)} \quad (41)$$

where ${}_2F_1(a, b, c; z)$ is the hypergeometric function.[10]

Let x_{thr} be 0.25. The probability $p(x > x_{thr})$ is computed using Mathematica for both slow and fast gene expression kinetics. The values of $p(x > x_{thr})$ in the slow and fast cases are 0.9294 and 0.9999 respectively. Since in the latter case, the probability that the protein level exceeds the threshold is higher, the chance of HI occurrence is correspondingly lower.

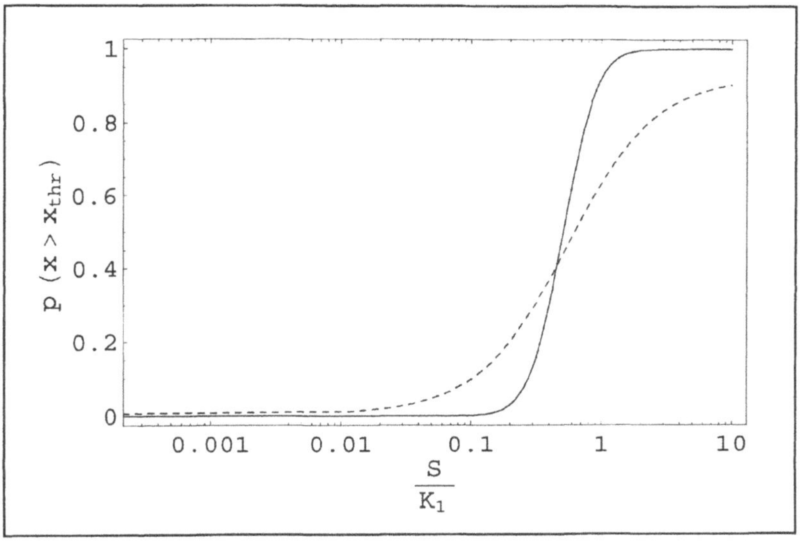

Figure 5. $p(x > x_{thr})$ versus S/K_1 in a semi-logarithm plot for $n = 1$. The solid (dotted) curve corresponds to fast (slow) kinetics.

In the presence of an inducing stimulus, say, TFs, the probability of activation above a threshold is again given by (40) with k_a and k_d replaced by k_a' and k_d' (Eq. (9)). Figure 5 shows $p(x > x_{thr})$, $x_{thr} = 0.25$, versus S/K_1 in a semi-logarithm plot for both slow and fast kinetics. In the fast case, a substantially steeper curve is obtained leading to enhanced signal discrimination, i.e., a more predictable response in a gradient of inducing signal. As shown by Cook et al,[3] the signal discrimination ability increases with gene copy number. One can thus speculate that diploid organisms utilise stochastic expression kinetics, preferably fast, for signal discrimination and are susceptible to degraded signal discrimination due to a reduction of gene copy number in the haploid state. Mutations in the subset of genes which generate a response to signaling gradients in diploid organisms may be the cause of some HI syndromes associated with these systems. Figure 6 shows the same plot as in Figure 5 but now the TF's form bound complexes with $n = 6$. In Eq. (9), $[S]$ is replaced by $[S_n] = K[S]^n$ (Eq. (12)). One now finds that the slopes of the curves for slow and fast kinetics are similar. Thus as n, the number of TF's forming the bound complex, increases, the distinction between slow and fast gene expression kinetics, as regards their signal discrimination ability, becomes less pronounced.

To sum up, in this section, we have described simple stochastic models of gene expression and shown that due to random fluctuations around the mean protein concentration in the steady state, the protein level may fall below the threshold even if this does not happen in the deterministic case. The variance, a measure of the spread around the mean protein level, is reduced with increasing gene copy number and faster expression kinetics. The variance increases if the rate constant j_p associated with protein synthesis is increased. In the case of one gene, we have further calculated the probability that the concentration of proteins exceeds a threshold in the absence as well as the presence of an inducing stimulus. In the latter case, faster gene expression kinetics give rise to a sharper response to changing stimulus concentrations. As shown by Cook et al,[3] this is also true when the gene copy number is increased. Thus the signal discrimination ability of diploid organisms may be impaired in the haploid state. When the inducing stimulus is a bound complex of n TFs, the distinction between slow and fast gene

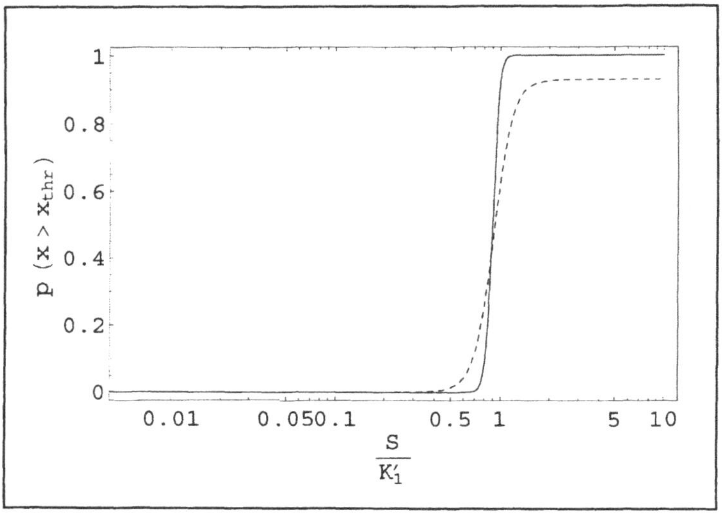

Figure 6. $p(x > x_{thr})$ versus S/K_1' $[(K_1')^n = K_1/K]$ in a semi-logarithm plot for $n = 6$. The solid (dotted) curve corresponds to fast (slow) kinetics.

expression kinetics becomes less with increasing *n*. In this Chapter, we have described both deterministic as well as stochastic models of gene expression and indicated possible scenarios for the occurrence of HI. Because of the simple nature of the models, a number of results can be derived analytically which provide a mathematical basis to the numerical simulation results of Cook et al[3] in the stochastic case. The sole ingredients of the minimal models described in this Chapter are universal processes like gene activation, deactivation and protein synthesis, each of which involves a number of biochemical events. The application of the simple models to the complex problem of HI, occurring in diploid organisms, is meant to illustrate the possible utility of such models in identifying the origins of HI.

Acknowledgement

R. K. is supported by the Council of Scientific and Industrial Research, India under Sanction No. 9/15 (239) / 2002 - EMR - 1.

References

1. Seidman JG, Seidman C. Transcription factor haploinsufficiency: when half a loaf is not enough. J Clin Invest 2002; 109:451-455
2. Veitia RA. Exploring the etiology of haploinsufficiency. BioEssays 2002; 24:175-184.
3. Cook DL, Gerber AN, Tapscott SJ. Modeling stochastic gene expression: implications for haploinsufficiency. Proc Natl Acad Sci 1998; 95:15641-15646.
4. McAdams HH, Arkin A. It's a noisy business! Genetic regulation at the nanomolar scale. Trends Genet 1999; 15:65-69.
5. Thattai M, van Oudenaarden A. Intrinsic noise in gene regulatory networks. Proc Natl Acad Sci 2001; 98:8614-8619.
6. Kepler TB, Elston TC. Stochasticity in transcriptional regulation: origins, consequences, and mathematical representations. Biophysical J 2001; 81:3116-3136.
7. Swain PS, Elowitz MB, Siggia ED. Intrinsic and extrinsic contributions to stochasticity in gene expression. Proc Natl Acad Sci 2002; 99:12795-12800.
8. Peccoud J, Ycart B. Markovian modeling of gene-product synthesis. Theor Pop Biol 1995; 48:222-234.
9. Van Kampen NG. Stochastic Processes in Physics and Chemistry. Amsterdam: North Holland, 1992.
10. Gradshteyn IS, Ryzhik IM. Table of Integrals, Series and Products: Academic Press, 1980.

CHAPTER 7

Biological Consequences of Dosage Dependent Gene Regulation in Multicellular Eukaryotes

James A. Birchler and Donald L. Auger

Abstract

Recent evidence from a variety of studies has indicated that gene regulatory mechanisms in multicellular eukaryotes operate in a dosage dependent manner. A consequence of this fact is that new mutations in regulatory loci are available for adaptive selection in the heterozygous state, which is consistent with previous hypotheses that morphological evolution occurs primarily via regulatory genes. Their dosage dependent nature also suggests a common basis for phenomena that result from changes in chromosomal dosage, such as aneuploid syndromes and dosage compensation. Moreover, many quantitative trait loci have been molecularly defined as regulatory genes that exhibit semi-dominant behavior, suggesting also a common basis in the dosage dependent nature of regulatory hierarchies.

Introduction

Recent data from several types of investigations have indicated that gene regulatory systems in multicellular eukaryotes operate in a dosage dependent manner. In other words, the quantity of individual components of regulatory complexes influences target gene expression in the diploid state and ultimately impacts the phenotype. Early evidence came from studies of gene expression in aneuploid individuals relative to the normal euploid. These studies indicated an extensive array of dosage sensitive effects on a single gene. Further realization of this concept came from analysis of developmental control genes and the hierarchies in which they participate. Many genes involved in early development typically operate in a concentration dependent fashion to condition the differential differentiation of various parts of the embryo. Yet another indication of this concept is the identification of the molecular basis of genes responsible for quantitative traits. Many of those elucidated to date exhibit semidominant behavior genetically and produce a dosage effect in those cases that have been tested. Lastly, the realization of the haplo-insufficiency of transcription factors for clinical manifestations in humans has added to the evidence. In this chapter we will review these areas of investigation and discuss the implications of the dosage dependent nature of regulatory gene interactions.

Gene Expression in Aneuploids

One of the first indications of the dosage dependent nature of eukaryotic regulatory hierarchies came from studies of trans-acting chromosomal dosage effects on the expression of

The Biology of Genetic Dominance, edited by Reiner A. Veitia. ©2006 Eurekah.com.

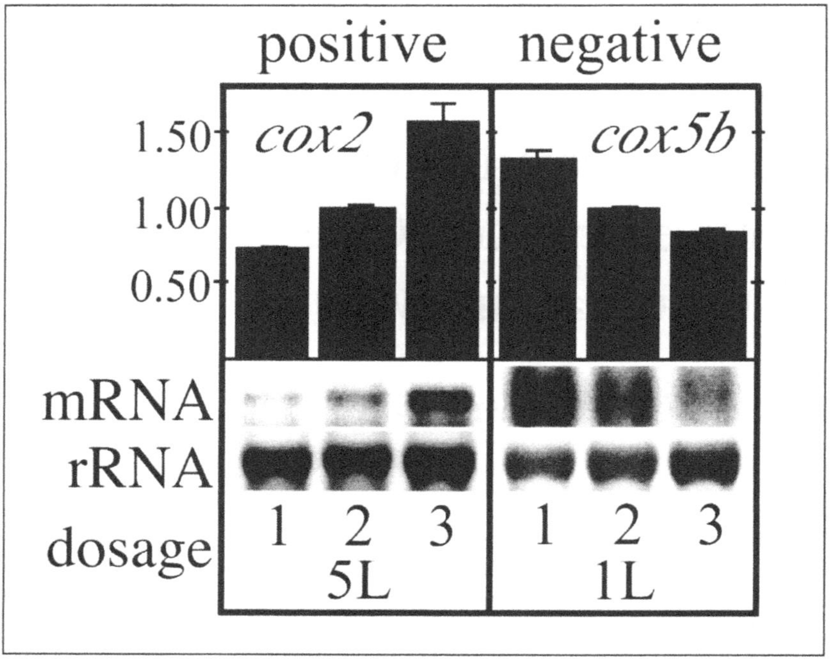

Figure 1. Examples of trans-acting dosage effects. A positive dosage effect is illustrated by the response of the *cox2* gene to the dosage of the long arm of chromosome 5 (5L) in maize. A negative dosage effect is shown by the response of the *cox5b* gene to the dosage of 1L. The mRNA level in each case was determined relative to the ribosomal RNA used as a Northern loading control. The source of the RNA was the embryo of maize kernels. Modified from Auger et al (2001).[6]

target genes throughout the genome. For example, when the long arm of chromosome 1 (1L) in maize was varied in a one to four dosage series, genes located on other chromosomes were modulated in expression.[1] Other chromosome arms produced similar effects.[2] The magnitude of these changes was within the range of an inverse correlation between the chromosomal dosage and the level of target gene expression. When the dosage of a chromosomal segment was reduced to one copy instead of the normal two, target genes were up-regulated within a two fold limit. However, when the same chromosomal segment was increased from the normal two copies to the trisomic state, target gene expression was reduced to two-thirds of the diploid. Further increase of the chromosomal dosage to four copies caused additional reductions of target gene expression to a lower limit of half the normal level.

In fewer cases, there was a positive correlation between chromosomal dosage and the amount of target gene expression. The monosomic situation caused reductions to nearly half of normal and the trisomic caused increases in gene expression up to 150%. Because the structural gene itself was not varied in copy number, this effect must have been due to an effect of regulatory loci.

Subsequent studies showed that these dosage effects operate on both the protein and RNA levels (Fig. 1).[3-7] It was found that any one gene product is sensitive to the trans-acting dosage effects of several but not all chromosomal segments. The effects operate on nuclear genes as well as on those encoded by the organelles.[6] Furthermore, the effects on gene expression caused by a particular chromosomal segment can be tissue specific.[8]

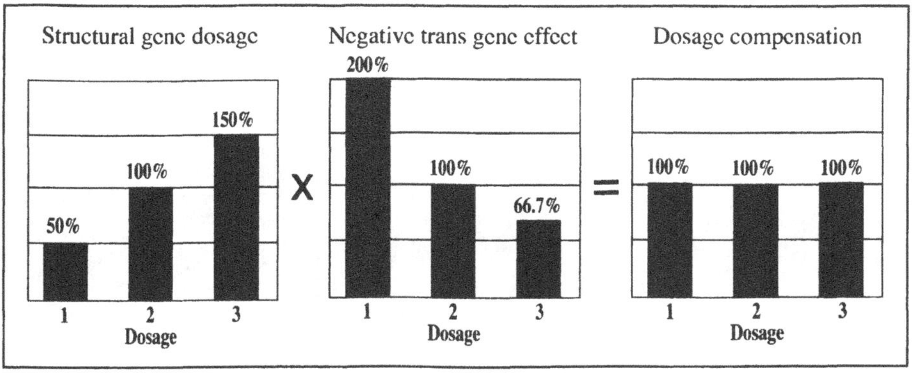

Figure 2. Dosage compensation. At the left is shown the gene expression typically exhibited by a structural gene in a one to three dosage series. In the center is shown the limits of the negative inverse dosage effect. Increases to a maximum of 200% of the normal are found in one dose individuals. Decreases to a lower limit of 67% are found in trisomics. When the two opposing effects occur in the same chromosome arm, they cancel to produce dosage compensation for the gene of interest as illustrated on the right.

The inverse or direct correlation limits are seldom breached despite the fact that a single gene product can be affected by multiple aneuploidies. This fact suggests that there is little cumulative effect of the various regulatory loci beyond these limits. This conclusion is further supported by the findings of gene expression studies in whole arm (20% of the genome) trisomics of Drosophila.[9] In these studies many gene products were assayed from around the genome. The overwhelming gene expression response to the trisomic condition was an inverse effect reducing expression to a lower limit of two-thirds of the normal level. Similarly, a phenotypic screen for dosage effects of the autosomes on the expression of the X linked *white* eye color gene found numerous regions that modulated *white* expression with the predominant effect being negative.[10]

Dosage Compensation

The observation that most trans-acting effects are negative helps to explain the fact that genes encoded on the varied segment often exhibit dosage compensation, i.e., the nearly equal expression of a target locus despite a change in dosage of the chromosome arm on which it resides (Fig. 2). In the original 1L dosage series in maize, the *Alcohol dehydrogenase1* gene showed nearly equal expression levels despite the fact that the structural gene was being varied. This compensation resulted from a cancellation of a gene dosage effect by a simultaneous inverse effect on *Adh1* produced by another 1L factor. When only a small region that included the *Adh1* structural gene was varied, a direct correlation between gene dosage and its expression was observed. However, when another region of 1L, which did not include *Adh1*, was varied from one to three doses, a negative effect on ADH levels was found.[11] In the whole of the chromosome arm, the two effects cancel each other to give dosage compensation.

Similarly, in the aforementioned whole arm trisomics of Drosophila, many genes on the varied arm exhibited compensation.[12] Even a segmental trisomic of substantial size surrounding the *Adh* gene in Drosophila was shown to exhibit dosage compensation for *Adh* RNA levels. This compensation was also due to a cancellation of the two types of opposing effects.[13]

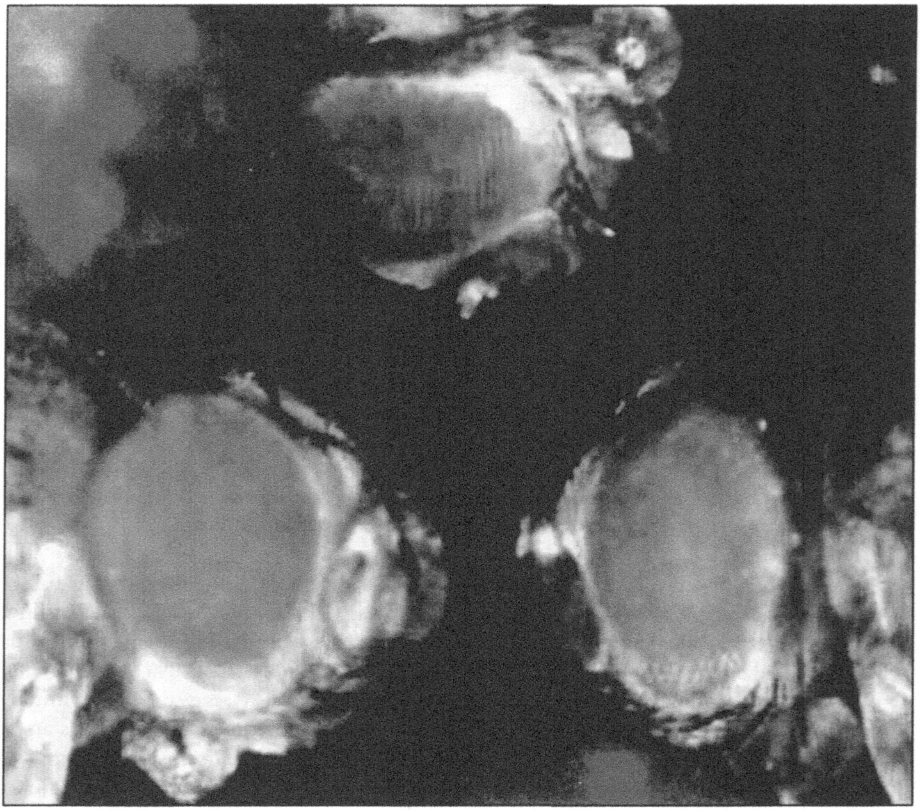

Figure 3. Inverse dosage effect of the X chromosome of Drosophila. A transgene of the *white* eye color gene was present in constant copy number in an autosome in a background in which the endogenous gene was deleted. When the dosage of the X chromosome was varied in one to three doses in the form of males, females and metafemales, the expression of the transgene was negatively correlated with the X dosage. Counterclockwise from the top, male, female, metafemale. See also Birchler (1992).[22]

Dosage Compensation of the X Chromosome in Drosophila

The same type of inverse dosage effect is exerted by the X chromosome of Drosophila.[14-16] The *white* eye color gene normally resides on the X chromosome and shows dosage compensation with the male having similar expression compared to females even though males have only one X.[17-21] A transgene of *white* was placed in the autosomes in a background in which the endogenous *white* gene was missing from its normal position. Variation of the X chromosome dosage from one to three in the form of males, females and metafemales (triple X) resulted in a negative effect on the expression of the *white* transgene, which was held at constant dosage (Fig. 3).[22] This result demonstrated that the X chromosome carries at least one negative trans-acting factor for *white*. Earlier results indicated that the *white* gene shows a gene dosage effect when only a small region surrounding it is varied.[17] Thus, on the normal X chromosome, the combination of a structural gene dosage effect and an inverse effect of regulatory genes acting simultaneously result in dosage compensation.

The dosage effects of the Drosophila X chromosome are more complicated than other chromosomes, probably because this situation has been exposed to natural selection in contrast to experimentally produced aneuploids.[23] Given that most trans-acting dosage effects are negative, one might expect that in males, where there is only one X, there would be a general elevation of autosomal gene expression.[24,25] Indeed, a tendency for male biased autosomal expression has recently been confirmed in microarray studies.[26] Nevertheless, the autosomes are not elevated to the extent that might otherwise be expected, because there is a specific mechanism to ameliorate the inverse dosage effect of the monosomic X upon the autosomes. This is accomplished by sequestrating histone acetylation complexes away from the autosomes in normal males.[27-30] The lower autosomal histone acetylation level is associated with lower transcriptional activity. If males carry mutations that eliminate the sequestration process, then the expression is now as predicted, i.e., a widespread up-regulation of autosomal gene expression due to the inverse dosage effect of the X chromosome.[31-34] In normal males, the complexes are sequestered to the single X chromosome. In order to prevent overcompensation, the complex overrides the impact of high levels of histone acetylation on the X itself, while still allowing the two-fold up-regulation for the proper level of dosage compensation.[32-34]

Regulatory Genes Are Responsible for the Inverse and Direct Dosage Effects

Individual genes that were responsible for the direct and inverse dosage effects were identified in mutagenic screens in Drosophila. The first case involved *Inverse regulator-a*.[35] When heterozygous with a null allele, there is a two-fold up-regulation of the *white* eye color gene, as well as other loci. The gene maps to a site in the second chromosome that, when normal alleles are varied, shows a negative dosage effect on *white* pigment and messenger RNA levels. A total of 47 dosage dependent modifiers of *white* have been identified (Table 1).[34] This compilation illustrates the number of dosage dependent factors that include a single target gene within their realm of influence, although they also have varying effects on other genes.

Studies involving multiples of these modifiers suggest that the effect of combinations is not cumulative beyond the inverse or direct correlation limits.[36] This result is consistent with the chromosomal dosage series. When nonallelic modifiers, each with an approximate two-fold up-regulatory effect on *white*, were combined, the response was never multiplicative, indicating that the various genes are not acting independently but rather are likely to function in a hierarchy of regulatory networks. Some pairs of modifiers showed epistasis in that the combined effect was more or less equivalent to that of one of the two modifiers.

It is noteworthy in that all characterized *white* modifiers are regulatory genes of one type or another. They include transcription factors, chromatin proteins, chromatin modifying enzymes and components of signal transduction cascades that are implicated in ultimately affecting gene transcription. The heterogeneous nature of this collection suggests that regulatory genes regardless of type tend to exhibit dosage effects.

Haplo-Insufficiency

The term haplo-insufficiency applies to the phenomenon in which detrimental effects are manifested for short monosomic regions. Recently, the term has been used to describe the fact that tumor suppressors and transcription factors produce a phenotype when present in only a single copy.[37] Previous thinking had held that either the uncovering of a mutant form of a tumor suppressor (loss of heterozygosity) or the simultaneous inactivation of both copies was necessary to condition cancer. Certainly these situations foster cancer development, but it has now been realized that hemizygosity of selected tumor suppressor genes also conditions the cancerous state, albeit at a slower rate.[38,39] It appears that a dosage effect of tumor suppressor genes is operating.

There is also a growing realization that a single functional copy of many transcription factors will produce human genetic conditions. Seidman and Seidman compiled a list of 32

Table 1. Dosage dependent modifiers of the white eye color gene of Drosophila

Modifier	Effect on *White*	Type of Regulator
abnormal, small, homeotic discs 2 (*ash2*)	negative	transcription factor
apterous (*ap*)	negative	LIM family transcription factor
Beadex (*Bd*)	negative	transcription factor
brahma (*brm*)	positive	chromatin remodeling
cap-n-collar (*cnc*)	negative	bZIP transcription factor
devenir (*dev*)	negative	trithorax-Group
Distalless (*Dll*)	negative	transcription factor
Enhancer of Polycomb (*E(Pc)*)	negative	Polycomb Group
Enhancer of white-spotted[81d5]	positive	unknown
Enhancer of zeste (*E(z)*)	negative	Polycomb Group
extra sex combs (*esc*)	negative	Polycomb Group
hedgehog (*hh*)	positive	cell-cell signaling
Inverse regulator-a (*Inr-a*)	negative	unknown
kismet (*kis*)	positive	trithorax-Group
kohtalo (*kto*)	negative	trithorax-Group
l(2) 05208/Kruppel-*homolog1* (*Kr-h1*)	positive	transcription factor
l(2) 03405/ Uba1	positive	Ubiquitin activating
l(3) 00305	negative	protein Ser/Thr kinase
l(3) 01969	negative	unknown
l(3) 02104	negative	unknown
l(3) 03670	negative	alkaline phosphatase
l(3) 04063	negative	unknown
l(3)04026	negative	unknown
l(3)08232 (rolling pebbles)	negative	apoptosis inhibitor ?
l(3)87Ca/Vha55	positive	Polycomb Group
Lightener of white (*Low*)	positive	unknown
mei-P19	negative	cell cycle control?
Modifier of white (*Mow*)	negative	unknown
modulo (*mod*)	negative	DNA binding
osa (*osa*)	positive	DNA binding
oxen (*ox*)	negative	diacyl glycerol kinase
Polycomblike (*Pcl*)	negative	Polycomb Group
Posterior sex combs (*Psc*)	negative	Polycomb Group
Regena (*Rga*)	negative	transcription factor
Ribonuclear protein at 97D	negative	nuclear RNA binding
Ribosomal protein PO/AP lyase (*PO*)	negative	AP3 DNA endonuclease
scalloped (*sd*)	negative	transcription factor
Sex combs on midleg (*Scm*)	negative	transcription factor
skuld (*skd*)	negative	trithorax-Group
sugarless (*sgl*)	negative	UDP-G6DH (signaling)
Suppressor of Polycomb (*Su(Pc)*)	negative	Polycomb Group
Trithoraxlike (*Trl*)	positive	transcription factor
Ultrafemale overexpression (*Ufo*)/ *lola*	negative	transcription factor
urdur (*urd*)	negative	trithorax-Group
verthandi (*vtd*)	negative	trithorax-Group
Weakener of white (*Wow*)	positive/negative	unknown
wingless (*wg*)	negative	cell-cell signaling

transcription factor loci that exhibit haplo-insufficient behavior for clinical manifestations.[40] These effects are likely the result of a dosage effect of the regulators on their target loci. The recognition of these dosage effects is consistent with work from other organisms and with the fact that aneuploidy produces severe developmental effects.[41-43]

What Is Dominance?

The question of dominance has been the focus of debate since Sewall Wright and Ronald Fisher introduced the issue in the late 1920s.[44-47] At the molecular level, dominance of an allele is likely to involve the presence of a biochemical function encoded by a housekeeping gene, while recessive alleles encode nonfunctional products or none at all.[48] Of target genes, only those that occupy the rate-limiting step of a biochemical pathway would produce a semi-dominant effect on the phenotype. The expression of the remaining target genes in a pathway would reflect the action of the regulatory system, whose components would exhibit a dosage effect and thus show semi-dominant behavior in heterozygotes.

The anthocyanin pigment pathway genes in maize provide an example. Most of the structural genes of the biochemical pathway have been identified.[49] Functional alleles show complete dominance over null alleles for five of the six known steps. The one exception is the *c2* gene that encodes the rate-limiting position in the pathway. In this case, a semi-dominant behavior is exhibited in heterozygotes. In contrast to the behavior of the target loci, the majority of the regulatory genes, namely *c1*, *r1*, and *in1*, show a semi-dominant behavior, at least under most conditions. Another regulator, the *vp1* gene, does not have a dosage effect on the phenotype. Thus, as in other cases, the majority of the regulatory genes are dosage dependent. Clearly, the degree of dominance can be quite variable, which contributes to a range of phenotypes.

How Could Multiple Regulators All Produce a Dosage Effect on the Same Target?

There are multiple regulatory genes that, when changed in dosage, alter the level of expression of any one target gene.[34,50-52] How could this occur mechanistically? The answer will require further study, but some data are available that shed light on this problem. For example, many helix-loop-helix transcription factors operate in multi-subunit complexes. One of the most thoroughly studied examples involves the regulators of the mammalian muscle differentiation pathway.[51] The stoichiometry of the individual transcription factors affects the activity of the complex as a whole. Thus, when the quantity of any one member is altered, the activity of the transcriptional complex is changed. In genetic terms, changing the dosage of a gene encoding any one component of a complex would alter the expression of the battery of target genes. Other examples include the Polycomb and Trithorax complexes of chromatin proteins.[53-55] Again, a complex exists composed of many subunits for which varying any individual member alters the effect of the whole.

The stoichiometric effects described above appear to explain why aneuploidy has a greater effect on target gene expression than do changes in whole genome ploidy.[2,13,56,57] The "per gene" expression among different ploidy levels usually is nearly equal.[57] The effects of aneuploidy are much more extensive, which fits with the concept that an altered stoichiometry of the regulatory genes causes these syndromes. An uncompensated gene dosage effect of some of the regulators on the varied segment relative to the remainder of the genome could produce this stoichiometric shift.

The second aspect of the regulatory mechanisms that helps explain the multiple factor observation is that hierarchies of dosage dependent regulators occur (Fig. 4). If dosage dependent regulators themselves are controlled by dosage dependent regulators, a system is in place such that variation in quantity of any one regulator in the hierarchy will modulate the ultimate

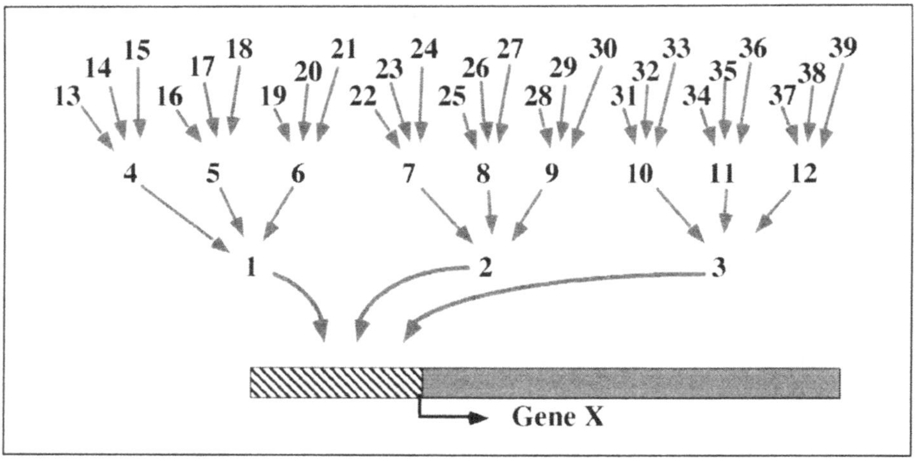

Figure 4. Heuristic model of a hierarchy of regulatory genes influencing a single target locus. If a single target gene is controlled by three dosage dependent regulators, which in turn are affected by the same number, only three levels of a hierarchy are needed for there to be 39 regulators potentially affecting the same target locus. The modulation in dosage of any one of the regulators in the hierarchy would in theory transmit a dosage effect to the ultimate target. Modified from Birchler et al (2001).[34]

target genes to some degree. A dosage effect at the top of a hierarchy will be transmitted through branching cascades to the targets. Studies of early development in Drosophila indicate that dosage dependent regulators control other transcription factors as targets, which in turn determine the amount and spectrum of the ultimate target genes expressed at any one time and place.[58-65] Such a system will contribute to the multi-factorial control of phenotypic traits.

Implications for Aneuploid Syndromes

As suggested above, the global effects on target gene expression observed in aneuploids are likely due to the altered stoichiometry in regulatory hierarchies.[2] A parallel relationship exists on the phenotypic level in that aneuploid syndromes are more severe than changes in ploidy (Fig. 5).[66] The global alterations of gene expression in aneuploids are both positive and negative and operate in both monosomics and trisomics.[2,5] Thus, with both increases and decreases of chromosomal dosage, many genes will exhibit lowered expression levels. The monosomic state has lower potential reductions due to those genes subject to positive dosage effects compared to the trisomic reductions for genes subject to the negative dosage effects. Both, of course, are reduced relative to the diploid. This correlation between the severity of the phenotype and the global patterns of gene expression suggest that when many target genes are altered in expression, the phenotype will be affected. There is likely to be a rate limiting aspect of gene expression on phenotypic characteristics and this fact would be reflected in aneuploid syndromes. Because the monosomic and trisomic conditions reduce many gene products, albeit different ones, both situations become rate limiting on the phenotype (Fig. 6).

Implications for the Genetic Basis of Quantitative Traits

Typically quantitative traits are controlled by multiple genes, whose alleles exhibit some degree of semi-dominant behavior.[67,68] In other words, alleles affecting quantitative traits often show dosage effects. This aspect of quantitative inheritance suggests a similar mechanism between its control and the dosage effects observed in aneuploids.[5] In both cases, multiple regions of the genome exhibit a dosage dependent influence on the phenotype.

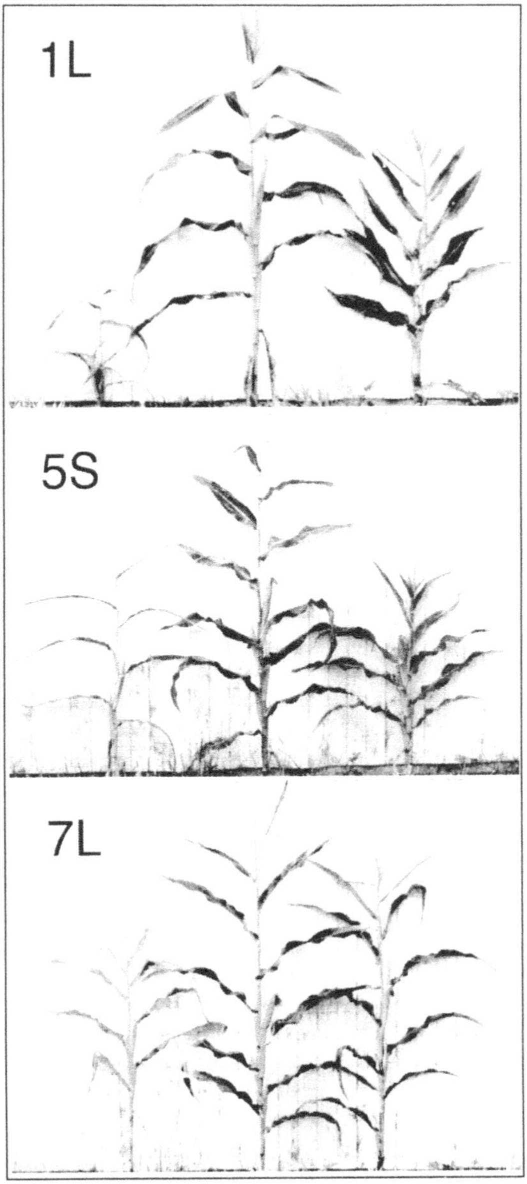

Figure 5. Aneuploid syndromes for different chromosome arms of maize. From left to right in each case are the monosomic, diploid and trisomic examples for chromosome arms, from top to bottom, 1L, 5S and 7L. Although each chromosome arm produces a characteristic syndrome, those produced by the monosomic condition tend to be more severe than those produced by the trisomics.

In recent years, some quantitative trait loci have been cloned and their molecular nature identified, primarily in plant species. Many of these QTL correspond to regulatory genes and, interestingly, usually exert a negative dosage effect. Examples include *teosinte branched* in maize, which controls plant morphology;[69] two genes in tomato affecting fruit size and shape;[70,71] the

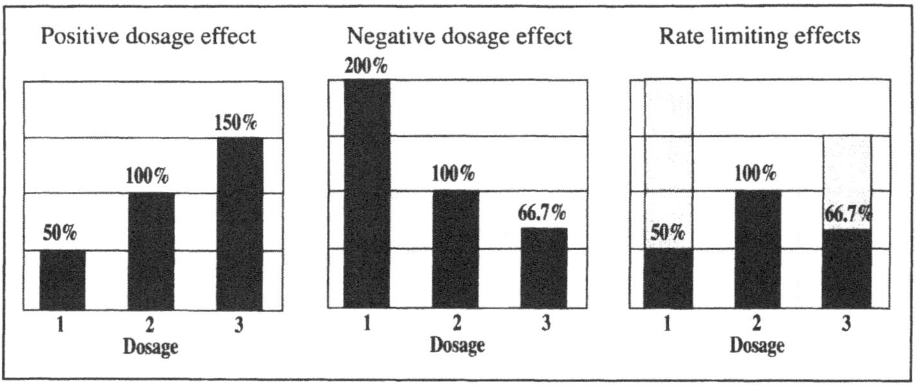

Figure 6. Rate limiting effects in aneuploids. At the right is illustrated the limits of changes in gene expression that result from positive dosage effects. In the center is the limits resulting from negative effects. Different sets of target loci are reduced in the monosomic and trisomic states to differing levels. If these reductions become rate limiting on growth, they could explain the typical aneuploid syndromes, which follow a similar pattern. Modified from Birchler et al (2001).[34]

Dwarf8 gene in maize, which affects stature and flowering time;[72] *FLC* in Arabidopsis, which affects flowering time;[73] the *p1* gene affecting maysin levels in maize;[74] and the *hairy* gene in flies affecting bristle number.[75] Thus, expression studies in aneuploids and the elucidation of the genes corresponding to quantitative trait loci suggest at least some common bases.

In addition, developmental processes are affected by regulatory hierarchies that also exhibit dosage effects. Variation in such loci no doubt would contribute to quantitative variation. These mechanisms would likewise be subject to selection at the diploid level. The use of gradients to formulate morphology is consistent with the dosage dependent nature of regulatory hierarchies.[58,59,65,76]

The Evolution of Regulatory Hierarchies in Multicellular Eukaryotes

In haploid organisms, gene expression levels can be selected directly in the haploid state. In contrast, in diploid organisms, mutations in regulatory processes that require homozygosity to modulate gene expression will not be accessible for natural selection to act under many circumstances. Only in small populations, in which random drift can fix rare mutations, would they become established in a population. Indeed the neutral theory of evolution suggests that adaptive evolution is rare and that most change occurs by drift.[77,78] This contention appears to be supported by the fact that comparisons of synonymous versus nonsynonymous base substitutions in protein encoding sequences are consistent with a low level of adaptive selection. Most of these studies, however, have concentrated on genes encoding metabolic or housekeeping functions rather than regulatory loci.[78,79] New mutations in metabolic target genes may be more likely to be neutral.

In contrast, the data showing that dosage effects are primarily due to regulatory genes suggest a loose dichotomy between regulatory genes and their target loci in how evolution acts upon them. It appears that regulatory systems in general have been selected to operate in such a manner that the quantity of individual molecules affects the phenotype in the diploid state. Allelic variation would be expected to show some level of semi-dominance. Newly arising mutations in regulatory genes would be immediately available for selection due to their impact in the heterozygote. Emerging evidence from analysis of regulatory factors supports the idea that they are more likely to be under adaptive rather than neutral evolution.[80-82]

It is also possible that there is balancing selection between alleles of a regulatory gene. The potential for pleiotropy is high for any allelic variant contributing to a regulatory complex that affects target gene expression in multiple tissues. An advantage conferred by an allele's effect on target gene expression in one tissue may be counteracted by a negative effect in another tissue or circumstance. The interplay of adaptive and balancing selection of regulatory variation has not been experimentally explored to any great extent, but is a subject that deserves attention.

Selfish and Cooperative Genes

The writings of George Williams, William Hamilton, Richard Dawkins and Mark Ridley present arguments that genes will evolve characteristics that foster their own perpetuation.[83-86] This idea is referred to as the "Selfish Gene" concept. While this term is often denigrated because genes have no motive, there should be little debate that characteristics aiding the survival of particular alleles in the gene pool will be selected. What is often forgotten in such discussions is that characteristics that develop "cooperation" among genes can be the most effective means of self-perpetuation. A system of gene regulation in which multi-component hierarchies are affected by a change in the quantity of any one member allows most regulatory genes the *potential* to contribute to the phenotype and reproductive success of the individual carrying them. In this manner, alleles of genes in the hierarchy will have the opportunity to foster their own perpetuation.

The evolution of such a multifactorial system also avoids the problem of too few regulatory genes contributing to the phenotype. If only a few genes contribute to the pool of variation, there might be too little available for survival as environmental conditions change. It would seem that those evolutionary lineages with a multifactorial control of phenotypic traits would be more likely to survive over evolutionary time.

Genomic Imprinting As a Nonmutational Means to Modulate Dosage Dependent Genes

Genomic imprinting refers to the phenomenon in which there is monoallelic expression resulting from the inactivation of one copy depending on the parental history. First described in maize for the behavior of the *r* locus expression in the endosperm,[87] there are now many cases known, particularly in mammals. One evolutionary rationale for parental imprinting involves parental conflict.[88] Females are postulated to evolve means to disperse resources equally among all progeny, whereas males would evolve mechanisms that foster more resources for their own progeny in competition with other potential fathers. While this concept can explain some facts, it does not appear to be consistent with others.[89]

Recently another explanation has been proposed to explain imprinting. This concept, referred to as the rheostat model, suggests that imprinting is a means of modulating dosage dependent genes that affect the phenotype.[90] An imprinting mechanism can change the expression level of a particular gene without causing a permanent mutation. This explanation fits well with the idea that regulatory hierarchies have evolved to be dosage dependent and provides an additional mechanism by which this can be achieved.

How Many Genes Are Important in Evolution?

There appears to be a loose dichotomy between regulatory genes and their targets in that the former typically show a dosage effect on the phenotype, while the latter do not. This would imply that when new mutations arise in a heterozygote, adaptive evolution is the major mode of evolution on regulatory processes and neutral evolution is the primary mode on the target genes. Positive selection will operate on new favorable alleles of regulatory genes and negative selection will occur against the less favorable alleles. Selection on gene regulatory mechanisms may be the major force in the evolution of multicellular diploid organisms. This concept was first proposed by King and Wilson,[91] following their comparison of protein profiles between chimp and human; the profiles were virtually indistinguishable despite the morphological differences between the two species (at least for most individuals).

This argument suggests that the number of genes driving the evolution of multicellular eukaryotes is dramatically smaller than the total. The fraction of the human or Drosophila genomes devoted to regulatory genes of various types approximates 10-15%.[92,93] Nevertheless, variation in these few thousand genes is sufficient to provide a unique genotype to each individual in a population of immense size. For example, with the 47 dosage dependent modifiers of the *white* gene noted above, if there were variation in each (unlikely, but for the sake of discussion) and these were independently assorting (again not true, but for the sake of discussion), then they would produce 140,737,488,835,533 genotypes for the expression level of *white*. To the extent that the expression levels of the various alleles are not the same and depending on the amount of epistasis, the number of phenotypic classes would approach a similar number. Clearly, this example is grossly oversimplified in terms of gene interactions and the level of variation present in populations, but it illustrates the concept that only a few thousand genes can produce an astounding number of possible genotypes and phenotypes.

No Need for "Hopeful Monsters"

Darwin envisioned gradualism in evolution to explain the changes in morphology and other characteristics of diverging lineages of organisms. In contrast Richard Goldschmidt proposed that dramatic changes could occur by single mutations that alter the characteristics of a species (Hopeful Monsters).[94] This suggestion was inspired by the rather significant alterations in developmental processes conditioned by so-called homeotic genes that are now known to control segment identity. The counter argument to Goldschmidt was that such large effect mutations are likely to be detrimental and rapidly selected against.

The realization that many, if not most, regulatory genes can exhibit dosage dependent behavior under at least some circumstances alleviates the need to postulate such Hopeful Monsters. The example of the 47 modifiers of *white* illustrates that, for even a simple characteristic, the potential exists for selection to change a phenotype dramatically through gradual steps, acting in series on different modifiers. Cryptic variation likely exists for quantitative traits in populations as illustrated in a study by Lauter and Doebley on the progenitor of domesticated maize, teosinte.[95] Hidden variation for certain morphological characteristics was present in teosinte populations that became evident only in a different genetic background.

Such cryptic variation may occur if certain regulatory genes are strongly rate limiting on a phenotype in one tissue, thus masking the variation in regulators that are less limiting on the same characteristic. The latter may be more strongly limiting on some other aspect of the total phenotype or on other target genes in the same tissue. In this manner, they would be maintained as dosage dependent. As alternative alleles of a strongly rate limiting regulator become fixed during selection, the rate limiting effects of other regulators in the hierarchy can become more subject to selection. Thus, by shifting which genes are limiting through the hierarchy, the phenotype can be pushed to greater extremes. Although the effect of any one variant allele is small, the cumulative effect of many genes of small effect changing the phenotype in one direction under selection can be much larger.

Hybrid Effects and Incompatibilities

When inbred lines are crossed together, the hybrid progeny in plants and animals are often more vigorous than either parent.[96,97] Particularly in plants, the hybrid exceeds the biomass of the better parent. This phenomenon is referred to as heterosis. In preliminary work from our laboratory, in which representative target genes were assayed for their level of expression, it appears that in the hybrid, "per cell" expression is often different from the mid-parent value of the two parents.[98] This result suggests that when allelic variation at regulatory loci is brought together in the same genome, novel interactions occur. A reasonable suggestion is that there are incompatibilities of allelic products that contribute to transcriptional regulatory complexes. Inasmuch as a majority of dosage dependent regulators of transcription act in a negative fashion, incompatibilities might lead to a global up-regulation of gene expression. At present it is unclear whether this is the case and what gene products might be responsible for heterosis.

Hybrids between species also often exhibit altered phenotypes not present in either parental species.[99-101] When analyzing the differences between *Drosophila simulans* and *mauritiana* for the size and shape of the male genitalia, which is the major morphological difference between the two species, there are multiple QTL.[102-104] For the most part, these genes exhibit additive or semi-dominant behavior. Thus, these loci appear to represent multiple dosage dependent genes—again showing the control of a specific trait by multiple regulators. In the case of interspecific crosses, sterility or lethality can also occur in hybrids. Some mutations have been identified that override these incompatibilities. To date, those that have been defined molecularly are dosage dependent regulatory genes—suggesting that the incompatibilities occur because of incongruous interactions in regulatory hierarchies.[80,105,106] The fact that the various loci function correctly in the respective species, but not when brought together in a hybrid, suggests that the various components of the regulatory hierarchy in any one lineage interact with each other and evolve accordingly to maintain functionality.

Concluding Remarks

No one would deny that global modulations of gene expression can affect the phenotype. Thus, an understanding of the mode of action and interaction of regulatory hierarchies in multicellular organisms is critical to appreciate many aspects of developmental processes and morphological evolution. The realization of the dosage dependent nature of many regulators is a step in that direction, but there is much to be learned about how regulatory complexes do in fact affect the phenotype. One aspect deserving further study involves the interactions of individual components of multimeric complexes. Also, a complete understanding of the impact of regulatory systems on the phenotype will require the elucidation of the pleiotropic effects of any one regulator throughout the organism. Another area deserving attention is how the epistatic interactions of genes contributing to various regulatory complexes operate. Lastly, understanding the means by which regulatory hierarchies are shifted and selected over time is an important goal for elucidating the evolution of regulatory mechanisms and the organisms they control.

Acknowledgements

Research in our laboratory has been supported by the US Department of Energy, National Science Foundation (USA) and the US Department of Agriculture. Nicole Riddle made many helpful suggestions on the manuscript.

References

1. Birchler JA. A study of enzyme activities in a dosage series of the long arm of chromosome one in maize. Genetics 1979; 92:1211-1229.
2. Birchler JA, Newton KJ. Modulation of protein levels in chromosomal dosage series of maize: The biochemical basis of aneuploid syndromes. Genetics 1981; 99:247-266.
3. Reichert GH. Two-dimensional gel analysis of proteins from mouse fetuses with trisomy 19 after DEAE-Sepharose chromatography. Genetical Research 1986; 47:193-197.
4. Klose J, Putz B. Analysis of two-dimensional protein patterns from mouse embryos with different trisomics. Proc Natl Acad Sci USA 1983; 80:3753-3757.

5. Guo M, Birchler JA. Trans-acting dosage effects on the expression of model gene systems in maize aneuploids. Science 1994; 266:1999-2002.
6. Auger DL, Newton KJ, Birchler JA. Nuclear gene dosage effects upon the expression of maize mitochondrial genes. Genetics 2001; 157:1711-1721.
7. Wanous MK, Munkvold JD, Kruse JD et al. Identification of chromosome arms influencing expression of the HMW glutenins in wheat. Theor Appl Genetics 2003; 106:213-220.
8. Cooper JL, Birchler JA. Developmental impact on trans-acting dosage effects in maize aneuploids. genesis 2001; 31:64-71.
9. Devlin RH, Holm DG, Grigliatti TA. The influence of whole-arm trisomy on gene expression in Drosophila. Genetics 1988; 118:87-101.
10. Sabl JF, Birchler JA. Dosage dependent modifiers of white alleles in Drosophila melanogaster. Genetical Research 1993; 62:15-22.
11. Birchler JA. The genetic basis of dosage compensation of alcohol dehydrogenase-1 in maize. Genetics 1981; 97:625-637.
12. Devlin RH, Holm DG, Grigliatti TA. Autosomal dosage compensation in Drosophila melanogaster strains trisomic for the left arm of chromosome 2. Proc Natl Acad Sci USA 1982; 79:1200-1204.
13. Birchler JA, Hiebert JC, Paigen K. Analysis of autosomal dosage compensation involving the Alcohol dehydrogenase locus in Drosophila melanogaster. Genetics 1990; 124:677-686.
14. Birchler JA, Hiebert JC, Krietzman M. Gene expression in adult metafemales of Drosophila melanogaster. Genetics 1989; 122:869-879.
15. Birchler JA. X chromosome dosage compensation in Drosophila. Science 1996; 272:1190.
16. Margolis OS. The effect of a supernumerary X chromosome on members of the Bar series of Drosophila. Genetics 1934; 19:18-84.
17. Muller HJ. Further studies on the nature and causes of gene mutations. Proc 6th Int Congr Genetics 1932; 1:213-255.
18. Muller HJ. Evidence of the precision of genetic adaptation. Harvey Lecture 1950; 43:165-229.
19. Stern C. Dosage compensation—development of a concept and new facts. Canadian Journal of Genetics and Cytology 1960; 2:105-118.
20. Baker BS, Gorman M, Marin I. Dosage compensation in Drosophila. Annu Rev Genetics 1994; 28:491-521.
21. Arkhipova I, Li J, Meselson M. On the mode of gene dosage compensation in Drosophila. Genetics 1997; 145:729-736.
22. Birchler JA. Expression of cis-regulatory mutants of the white locus in metafemales of Drosophila melanogaster. Genetical Research 1992; 59:11-18.
23. Birchler JA. Inverse effect regions in maize and Drosophila and their possible role in dosage compensation, sexual dimorphism of autosomal genes and sex determination in the latter. Maize Genetics Cooperation News Letter 1977; 51:18-22.
24. Smith PD, Lucchesi JC. The role of sexuality in dosage compensation in Drosophila. Genetics 1969; 61:607-618.
25. Birchler JA. Genetic analysis of a modifier of the sexual dimorphism of glass in Drosophila melanogaster. Genetical Research 1984; 44:125-132.
26. Parisi M, Nuttall R, Naiman D et al. Paucity of genes on the Drosophila X chromosome showing male-biased expression. Science 2003; 299:697-700.
27. Kuroda MI, Kernan M, Kreber R et al. The maleless protein associates with the X chromosome to regulate dosage compensation in Drosophila. Cell 1991; 66:935-947.
28. Bone JR, Lavender RJ, Richman R et al. Acetylated histone H4 on the male X chromosome is associated with dosage compensation in Drosophila. Genes and Development 1994; 8:96-104.
29. Hilfiker AD, Hilfiker-Kleiner D, Pannuti A et al. mof, a putative acetyl transferase gene related to the Tip60 and MOZ human genes and to the SAS genes of yeast, is required for dosage compensation in Drosophila. EMBO J 1997; 16:2054-2060.
30. Turner BM, Birley AJ, Lavender J. Histone H4 isoforms acetylated at specific lysine residues define individual chromosomes and chromatin domains in Drosophila polytene nuclei. Cell 1992; 69:376-384.
31. Hiebert JC, Birchler JA. Effects of the maleless mutation on X and autosomal gene expression in Drosophila melanogaster. Genetics 1994; 136:913-926.

32. Bhadra U, Pal-Bhadra M, Birchler JA. Role of the male specific lethal (msl) genes in modifying the effects of sex chromosomal dosage in Drosophila. Genetics 1999; 152:249-268.
33. Bhadra U, Pal Bhadra M, Birchler JA. Histone acetylation and gene expression analysis of Sex lethal mutants in Drosophila. Genetics 2000; 155:753-763.
34. Birchler JA, Bhadra U, Pal Bhadra M et al. Dosage dependent gene regulation in multicellular eukaryotes: Implications for dosage compensation, aneuploid syndromes and quantitative traits. Developmental Biology 2001; 234:275-288.
35. Rabinow L, Nguyen-Huynh AT, Birchler JA. A trans-acting regulatory gene that inversely affects the expression of the white, brown and scarlet loci in Drosophila melanogaster. Genetics 1991; 129:463-480.
36. Bhadra U, Pal Bhadra M, Birchler JA. Interactions among dosage-dependent trans-acting modifiers of gene expression and position-effect variegation in Drosophila. Genetics 1998; 150:251-263.
37. Veitia RA. Exploring the etiology of haploinsufficiency. Bioessays 2002; 24:175-184.
38. Goss KH, Risinger MA, Kordich JJ et al. Enhanced tumor formation in mice heterozygous for Blm mutation. Science 2002; 297:2051-2053.
39. Fodde R, Smits R. A matter of dosage. Science 2002; 298:761-763.
40. Seidman JG, Seidman C. Transcription factor haploinsufficiency: When half a loaf is not enough. J Clin Invest 2002; 109:451-455.
41. Epstein CF. The Consequences of Chromosome Imbalance, Principles, Mechanisms and Models. Cambridge University Press, 1986.
42. Lindsley DL, Sandler L et al. Segmental aneuploidy and the genetic gross structure of the Drosophila genome. Genetics 1972; 71:157-184.
43. Bond DJ, Chandley AC. Aneuploidy. Oxford University Press, 1983.
44. Fisher RA. The possible modifications of the response of the wild type to recurrent mutations. Amer Nat 1928; 62:115-126.
45. Fisher RA. Two further notes on the origin of dominance. Amer Nat 1928; 62:571-574.
46. Wright S. Fisher's theory of dominance. Amer Natl 1929; 63:274-279.
47. Wright S. Physiological and evolutionary theories of dominance. Amer Natl 1934; 68:24-53.
48. Orr HA. A test of Fisher's theory of dominance. Proc Natl Acad Sci USA 1991; 88:11413-11415.
49. Dooner HK, Robbins TP, Jorgensen RA. Genetic and developmental control of anthocyanin biosynthesis. Ann Rev Genetics 1991; 25:173-199.
50. Henikoff S. Dosage-dependent modification of position-effect variegation in Drosophila. Bioesssays 1996; 18:401-409.
51. Weintraub H. The myoD family and myogenesis: Redundancy, networks and thresholds. Cell 1993; 75:1241-1244.
52. Grandori C, Cowley SM, James LP et al. The MYC/MAX/MAD network and the transcriptional control of cell behavior. Annu Rev Cell Dev Biol 2000; 16:653-699.
53. Kennison JA, Russell MA. Dosage dependent modifiers of homoeotic mutations in Drosophila melanogaster. Genetics 1987; 116:75-86.
54. Saurin AJ, Shao Z, Erdjument-Bromage H et al. A Drosophila Polycomb group complex includes Zeste and dTAFII proteins. Nature 2001; 412:655-660.
55. Bel S, Core N, Djabali M et al. Genetic interactions and dosage effects of Polycomb group genes in mice. Development 1998; 125:3543-3551.
56. Lucchesi JC, Rawls Jr JM. Regulation of gene function: A comparison of enzyme activity levels in relation to gene dosage in diploids and triploids of Drosophila melanogaster. Biochem Genetics 1973; 9:41-51.
57. Guo M, Davis D, Birchler JA. Dosage effects on gene expression in a maize ploidy series. Genetics 1996; 142:1349-1355.
58. Driever W, Nusslein-Volhard C. The bicoid protein determines position in the Drosophila embryo in a concentration-dependent manner. Cell 1988; 54:95-104.
59. Struhl G, Struhl K, Mcdonald PM. The gradient morphogen bicoid is a concentration-dependent transcriptional activator. Cell 1989; 57:1259-1273.
60. Warrior R, Levine M. Dose-dependent regulation of pair-rule stripes by gap proteins and the initiation of segment polarity. Development 1990; 110:759-767.

61. Sauer F, Jackle H. Concentration-dependent transcriptional activation or repression by Kruppel from a single binding site. Nature 1991; 353:563-566.
62. Cribb DL, Benassayag C, Randazzo FM et al. Levels of homeotic protein function can determine developmental identity: Evidence from low-level expression of the Drosophila homeotic gene proboscipedia under Hsp70 control. EMBO J 1995; 14:767-778.
63. Schulz C, Tautz D. Autonomous concentration-dependent activation and repression of Kruppel by hunchback in the Drosophila embryo. Development 1994; 120:3043-3049.
64. Abrell S, Carrera P, Jackle H. A modifier screen of ectopic Kruppel activity identifies autosomal Drosophila chromosomal sites and genes required for normal eye development. Chromosoma 2000; 109:334-342.
65. Christian JL. BMP, Wnt and Hedgehog signals: How far can they go? Curr Opin Cell Biol 2000; 12:244-249.
66. Blakeslee AF, Belling J, Farnham ME. Chromosomal duplication and Mendelian phenomena in Datura mutants. Science 1920; 52:388-390.
67. Mackay TFC. The genetic basis of quantitative variation: Numbers of sensory bristles of Drosophila melanogaster as a model system. Trends in Genetics 1995; 11:464-470.
68. Tanksley SD. Mapping polygenes. Annu Rev Genetics 1993; 27:205-233.
69. Doebley JF, Stec A, Hubbard L. The evolution of apical dominance in maize. Nature 1997; 386:485-488.
70. Frary A, Nesbitt RC, Frary A et al. fw2.2: A quantitative trait locus key to the evolution of tomato fruit size. Science 2000; 289:85-88.
71. Liu J, Van Eck J, Cong B et al. A new class of regulatory genes underlying the cause of pear-shaped tomato fruit. Proc Natl Acad Sci USA 2002; 99:13302-13306.
72. Thornsberry JM, Goodman MM, Doebley J et al. 4th. Dwarf8 polymorphisms associate with variation in flowering time. Nature Genetics 2001; 28:286-289.
73. Michaels SD, Amasino RM. Flowering Locus C encodes a novel MADS domain protein that acts as a repressor of flowering. Plant Cell 1999; 11:949-956.
74. Byrne P, McMullen M, Snook ME et al. Quantitative trait loci and metabolic pathways: Genetic control of the concentration of maysin, a corn earworm resistance factor, in maize silks. Proc Natl Acad Sci USA 1996; 93:8820-8825.
75. Robin C, Lyman RF, Long AD et al. hairy: A quantitative trait locus for Drosophila sensory bristle number. Genetics 2002; 162:155-164.
76. Jiang J, Levine M. Binding affinities and cooperative interactions with bHLH activities delimit threshold responses to the dorsal gradient morphogen. Cell 1993; 72:741-752.
77. Kimura M. Evolutionary rate at the molecular level. Nature 1968; 217:624-626.
78. Fay JC, Wu C-I. The neutral theory in the genomic era. Curr Opin Genes Develop 2001; 11:642-646.
79. Sabeti PC, Reich DE, Higgins JM et al. Detecting recent positive selection in the human genome from haplotype structure. Nature 2002; 419:832-837.
80. Barbash DA, Siino DF, Tarone AM. A rapidly evolving MYB-related protein causes species isolation in Drosophila. Proc Natl Acad Sci (USA) 2003; in the press.
81. Barrier M, Robichaux RH, Purugganan MD. Accelerated regulatory gene evolution in an adaptive radiation. Proc Natl Acad Sci USA 2001; 98:10208-10213.
82. Barrier M, Bustamante CD, Yu J et al. Selection on rapidly evolving proteins in the Arabidopsis genome. Genetics 2003; 163:723-733.
83. Williams GC. Adaptation and Natural Selection. Princeton. Princeton, NJ: University Press, 1966.
84. Hamilton WD. Narrow roads of geneland. Oxford: W. H. Freeman, 1996.
85. Dawkins R. The selfish gene. Oxford, UK: Oxford University Press, 1976.
86. Ridley M. The cooperative gene. Simon and Schuster, New York: The Free Press, 2001.
87. Kermicle JL. Dependence of the R-mottled aleurone phenotype in maize on mode of sexual transmission. Genetics 1970; 66:69-85.
88. Moore T, Haig D. Genomic imprinting in mammalian development: A parental tug-of-war. Trends in Genetics 1991; 7:45-49.
89. McVean GT, Hurst LD. Molecular evolution of imprinted genes: No evidence for antagonistic coevolution. Proc R Soc Lond Biol Sci 1997: 264:739-746.

90. Beaudet AL, Jiang Y-h. A rheostat model for a rapid and reversible form of imprinting-dependent evolution. Am J Hum Genet 2002; 70:1389-1397.
91. King MC, Wilson AC. Evolution at two levels in humans and chimpanzees. Science 1975; 188:107-116.
92. Adams MD, Celniker SE, Holt RA et al. The Genome Sequence of Drosophila melanogaster. Science 2000; 287:2185-2195.
93. Venter JC, Adams MD, Myers EW et al. The Sequence of the Human Genome. Science 2001; 291:1304-1351.
94. Goldschmidt RB. Theoretical Genetics. Berkeley, California: University of California Press, 1955.
95. Lauter N, Doebley J. Genetic variation for phenotypically invariant traits detected in teosinte: Implications for the evolution of novel form. Genetics 2002; 160:333-342.
96. Bruce AB. The Mendelian theory of heredity and the augmentation of vigor. Science 1910; 32:627-628.
97. Coors JG, Pandey S. Genetics and Exploitation of Heterosis in Crops. Madison, Wisconsin: American Society of Agronomy, Inc. and Crop Science Society of America, Inc., 1999.
98. Osborn TC, Pires JC, Birchler JA et al. Understanding mechanisms of novel gene expression in polyploids. Trends in Genetics 2003; 19:141-147.
99. Dobzhansky T. Genetics and the Origin of Species. New York: Columbia University Press, 1937.
100. Muller HJ. Isolating mechanisms, evolution and temperature. Biol Symp 1942; 6:71-125.
101. Turelli M, Orr HA. Dominance, epistasis and the genetics of postzygotic isolation. Genetics 2000; 154:1663-1679.
102. Liu J, Mercer JM, Stam LF et al. Genetic analysis of a morphological shape difference in the male genitalia of Drosophila simulans and D. mauritiana. Genetics 1996; 142:1129-1145.
103. True JR, Weir BS, Laurie CC. A genome-wide survey of hybrid incompatibility factors by the introgression of marked segments of Drosophila mauritiana chromosomes into Drosophila simulans. Genetics 1996; 142: 819-837.
104. Laurie CC, True JR, Liu J et al. An introgression analysis of quantitative trait loci that contribute to a morphological difference between Drosophila simulans and D. mauritiana. Genetics 1997; 145:339-348.
105. Ting C-T, Tsaur S-C, Wu M-L et al. A rapidly evolving homeobox at the site of a hybrid sterility gene. Science 1998; 282:1501-1504.
106. Peres DE, Wu C-I. Further characterization of the Odysseus locus of hybrid sterility in Drosophila: One gene is not enough. Genetics 1993; 140:201-206.

CHAPTER 8

Clusters of Functionally Related Genes in Eukaryotes, Dosage Balance and Evolvability

Reiner A. Veitia

Abstract

There is growing genomic evidence on clustering of functionally related genes in eukaryotes. Recent studies in yeast show a correlation between the expression patterns of adjacent and nonadjacent pairs of genes and that a significant proportion of the former are functionally related. A transcriptome-interactome correlation has provided global evidence that yeast genes with similar expression profiles are more likely to encode interacting proteins. Finally, an analysis of several sequenced eukaryotic genomes found that genes involved in the same pathways were clustered. Stoichiometric imbalances in macromolecular complexes can be a source of dominant phenotypes. Here this "dosage balance hypothesis" is extended with the idea that in many instances the physical association of eukaryotic genes might result from a mechanism selecting "modules" which ensure better coregulation, maintain the right stoichiometry of complexes and facilitate evolution by gene duplication.

Introduction

Clustering of genes is widespread in prokaryotes due to the existence of operons. Bacterial operons are sets of cotranscribed genes that participate in the same or related functions. In bacteria, operons are expressed as polycistronic transcripts. Surveys across sequenced bacterial genomes show that few operons remain intact over long periods of time and that the most stable gene clusters within the operons are those encoding products that physically interact.[1] It is interesting to put this fact in the perspective of the ideas prompted to explain the existence of operons. Notably, one of these hypotheses proposes that operon organization allows the production of locally concentrated proteins which facilitate their interactions (physical or metabolic). The molarity hypothesis relies on the fact that transcription and translation are coupled processes in prokaryotes.[2]

Gene arrangement in operons essentially lacks in eukaryotes, although they have been documented in *Caenorhabditis elegans*.[3] In fact, in the nematode about 1/4 of the genes are organized in special operons leading to polycistroninc transcripts which are then resolved by both trans- or cis-splicing into monocistronic messages suitable for translation. Interesting for the discussion that will follow, these operons usually contain genes encoding several subunits of macromolecular complexes.[3,4]

In most eukaryotes, trans-regulatory factors are supposed to ensure coregulation of genes involved in the same pathway. This is the general rationale to explain the almost universal absence of operon organization in eukarya. In spite of this, the existence of post-transcriptional eukaryotic operons exploiting the combinatorial use of nontranslated signals in the mRNAs

The Biology of Genetic Dominance, edited by Reiner A. Veitia. ©2006 Eurekah.com.

has been suggested.[5] Again this idea, that seems very likely, has a translation in terms of a dosage balance of the products encoded by the genes involved in the post-transcriptional operon.

Several examples of clusters of functionally related genes in eukaryotes have been known for a long time (i.e., gene families that have arisen by gene duplication and divergence, as globins or homeobox genes). These instances where the evolutionary pathway is clear and that involve structurally related genes (ancient paralogs) will not be dealt with here. Other, more interesting examples, are available but they have been considered as exceptions or doubted owing to the anecdotal nature of the reports (see refs. 6, 7). Today, there is compelling genomic evidence pointing to the existence of clusters of functionally related genes that goes beyond these isolated cases. For instance, a systematic analysis of microarray data in yeast showed a statistical correlation between the expression patterns of adjacent, as well as nearby nonadjacent pairs of genes. A statistically significant proportion of adjacent pairs examined fell into the same MIPS functional category (MIPS: Munich Information Center for Protein Sequences).[8] Later, a transcriptome-interactome correlation mapping using yeast data sets provided global evidence that genes with similar expression profiles were more likely to encode interacting proteins.[9] Lercher et al (2001)[10] have reported that human housekeeping genes (i.e., expressed in most tissues) show strong clustering. On top, in a recent work, Lee and Sonnhammer (2003)[11] have analyzed five sequenced eukaryotic genomes looking for clustering of genes assigned to the same pathway in the KEGG database (KEEG: Kyoto Encyclopedia of Genes and Genomes). They have found that many of the analyzed pathways exhibited significantly higher clustering than expected by chance. Virtually all pathways in yeast showed significant clustering followed by man, *Caenorhabditis*, *Arabidopsis*, and *Drosophila*. Only 7 out of 69 pathways found in all species were significantly clustered in all of them. The authors attributed the species-specific pattern of pathway clustering to adaptations or evolutionary events unique to a particular lineage.[11]

Physical association of genes encoding functionally related genes could have a trivial explanation if genes were controlled by limiting amounts of transcription factors. Here, "limiting" does not translate automatically into low concentration but into low availability due to the crowded nature of the nucleoplasm where there must be many nonspecific interactions. In these cases, diffusion times would be reduced if genes undergoing the same regulation were confined to specific chromatin territories (as conceived in ref. 12). This would be favored by their short-range physical association. However, recent in vivo imaging work suggests that the nucleoplasm is a highly dynamic medium concerning diffusion, as shown for many subunits of the RNA pol I and its transcription complexes.[13] In addition, if diffusion were a problem, the first solution would have been to remove the vast amount of junk DNA that is stored in the nucleus. One may also think that clustering functionally related genes into the same chromatin domains (i.e., under the action of a locus control region or enhancer) is advantageous. However, taking into account that diffusion does not seem to be a problem, it is not clear why controlling one cluster would be more efficient than regulating dispersed components. Surely, the cost of unfolding chromatin implies a weak selective pressure (one regulation event if clustered versus multiple if not). But if this cost is crucial, why is clustering observed for several but not all the actors of a pathway? And why only (according to ref. 10) for housekeeping genes that are ubiquitously expressed from a chromatin that may retain covalent marks to remain in an open conformation?

Another plausible explanation that is worth considering in the present context is related to the impact of stoichiometric changes on macromolecular assembly. Following the specific topology of a complex, increased or decreased amounts of one component can diminish the yield of the whole complex (titrating effect) and lead to an abnormal phenotype.[14,15] As already noticed above in the discussion about the operons, it seems that a certain dosage balance among the components of a macromolecular complex must exist to ensure a normal phenotype. Very recently, the dosage balance theory has been tested in yeast.[16] In short, the authors examined the heterozygous fitnesses of yeast knockouts of essential genes (i.e., homozygous fitness = 0) and data on artificial gene over-expression. They find that both under and over-expression of protein complex subunits lower fitness more than for genes that are not involved in complexes. Crucial in the current context, they also find that the accuracy of transcriptional coregulation of subunits is proportional to the deleterious consequences of imbalance, that is, when the degree of coexpression is higher, the heterozygote fitness is lower. The dosage balance theory predicts that single duplication of gene encoding certain types of protein subunits is expected to be harmful, as this may lead to imbalance. As expected, they find that members of large gene families are rarely involved in complexes. As a natural extension of the balance theory, I propose that physical association of genes coding for certain complex components and their partners is a selected property that ensures coregulation, and maintains the right stoichiometry of complexes which is advantageous in cases of partial aneuploidy. This facilitates evolution by gene duplication. For the sake of generality, other "linked configurations" directly related to dosage compensation will also be discussed in a later section. The reader must keep in mind that the aim of this essay is to outline one plausible explanation about the formation of clusters of interacting genes and not to provide the universal explanation of this fact.

Some Theoretical Considerations

Components of a macromolecular complex can be classified from a topological point of view, (and without being exhaustive) into several types.[15,17] (i) When a sub-unit forms a single bridge (as B in A-B-C) between two or more separable parts (A and C), a modest increase in the concentration of the former may induce a strong inhibition of complex formation due to existence of "kinetic traps". A kinetic trap is a point of no return during complex formation: for instance, excess of B in A-B-C leads to a proportion of inactive subcomplexes AB and BC in irreversible conditions. Monomer B is said to be titrating.[14] (ii) When momoners form multiple bonds within the body of the complex but without forming a link between separable parts, increased concentrations lead to abnormal intermediates that can eventually close up into the definitive complex in reversible conditions. In irreversibility, this class of components can also lead to inactive complexes. (iii) An increase in the concentration of a sub-unit linked to only one type of structure through a single bond does not show any negative effect. However, (iv) the latter monomers are identical (as in A-B-A) a decrease in their concentration may lead to a nonproportional reduction in complex formation, especially during irreversible reactions.[15] Besides, (v) if the monomers can form alternative complexes (i.e., AA, AB and BB) any subunit imbalance will affect relative complex concentration. Note here that the previous classification is general and not restricted to protein complexes.

The statistical evidence provided for the dosage balance hypothesis[16] suggests that, at least in yeast, many protein complexes form under essentially irreversible conditions and that imbalance might lead to inactive complexes. This is perhaps due to the very short generation time of this organism, which requires association reactions to take place promptly. Consider in this context that most protein-protein interactions in solution are not far from being diffusion-limited with specific association rates (*ka*) of the order of 10^6 to 10^8 $M^{-1}s^{-1}$. A diffusion-limited reaction takes place at the highest rate allowed by the viscosity of the medium. Thus, the wide ranges in the association constants observed in vitro depend mostly on the dissociation rates (*kd*) of the complexes ($10^3 s^{-1}$ to 0.25 months^{-1}, remember that $K=ka/kd$).[18] For diffusion-limited assembly reactions, the right stoichiometry among the components would be advantageous and particularly for those with very low *kd*.

There are many possible physicochemical strategies to attain and preserve the right stoichiometry in a complex or palliate its alterations.[15] The cell may exploit cooperativity and sequential assembly through routes that may differ from one organism to another. This plasticity can be studied with computational experiments where sets of equilibrium constants are "trained" to give a desired dose-response curve by a process analogous to evolution.[17] It is intuitively clear that the same dose-response curve can be obtained with many different sets of parameters. This is, indeed, the molecular basis of compensatory mutations. This translates into the fact that in the population individuals with different genotypes will be able to generate the same biological responses. The population will therefore harbor a wide neutral genetic variation.[19]

A way to preserve stoichiometry is to select chromosomal configurations that keep dosage sensitive genes closely linked to ensure a better coregulation. Interestingly, computer simulations show that when a strongly titrating factor (type i) is coincreased with a separable component directly linked to it, the titrating power diminishes in irreversible reactions. For instance, suppose that to form complex A-B-C, the initial relative amounts/rate of synthesis of A, B and C are as in the formula and that all specific rate constants are identical (other possibilities of starting concentrations and kinetic constants lead to similar results). As mentioned above, an excess of B dramatically interferes with ABC formation. However, a parallel increase (even a duplication) of B and another partner (A or C) leads to a normal amount of trimer (100%). Moreover, a parallel halving of both monomers does not have a higher effect than deleting one of them alone (50% normal). The titrating power of B is much higher in a complex ABCD where A, C and D are separable components (for instance, 1.5X B may lead to over 60% reduction of ABCD). In this case, coincrease of B with one component diminishes titration (yield ABCD > 65%) and coincrease with two components virtually removes the problem. Again, diminution of *B* and one or two of its syntenic partners is not worse than diminishing one of them alone. The titrating property of B also appears in a complex A-B-A. Increasing *A* has no effect, as expected for a nontitrating component, but an increase of *B* diminishes the yield of ABA (i.e., ~70% instead of 100%). It is clear that gene clustering would also be advantageous here. This applies to deletions: deletion of one copy of gene *A* alone leads to ~30% of complex yield instead of 50%.[15] However, heterozygous deletions of both *A* and *B* lead to a 50% reduction of trimer yield. In bigger complexes, AnB (n>2), the nonlinear effects of either increasing B or halving A sharpen as n increases. The relevance of clustering is also obvious for genes *A* and *B* whose products form homo- and heterodimers with different functions that must respect the relative molar concentrations. For instance, under assumptions as above (equal starting concentrations and specific rates), normally we have 1AA:2AB:1BB. However, halving B alone leads to 4AA:4AB:1BB while trisomy of B yields 1AA:3AB:2.25BB. The importance of clustering of components of type (ii) which form multiple bonds within the body of the complex but without linking separable parts is less obvious to study due to the immense space of topological possibilities. However, it is certain that examples of titration by this type of components can be found, especially in conditions of irreversibility, but this is beyond the nonexhaustive scope of this discussion. A property that emerges from the view presented here is that physical association of genes is relevant to avoid imbalance during duplication events but a side effect of this is to palliate imbalance in some cases of deletion.

Is Clustering of Interacting Genes Conceivable?

Physical clustering as outlined above is beneficial to palliate the immediate effect of aneuploidy. However, clustering cannot be conceived as anticipatory to overcome the deleterious effects of unscheduled genome rearrangements. Selection may act to establish and keep a new gene order and result in clustering of genes that are involved in the same complex. To illustrate this, let us briefly consider the evolutionary history of a "supergene". A supergene is a set of linked genes that are transmitted as a unit. The most common view of the formation of a supergene involves a mechanism that hinders recombination (i.e., a chromosomal inversion).

Thus, allele combinations of the loci included in the nonrecombining region tends to be transmitted together.[20] Here, I present a slightly more complex version of the above to explain how a tight cluster of interacting loci can arise. Following population genetic arguments prompted by Bodmer in 1962,[6] let us focus on a sexual diploid population where genes *A* and *C* encode interacting partners (one of both is highly dosage sensitive) and are arrayed as *ABC*

In the most stringent case, assume that alleles *A* and *C* are frequent while *a* and *c* are rare but their combination (haplotype *ac*) confers an advantage due, for instance, to a better coregulation. Individuals *AC/AC* and *AC/ac* will have fitnesses in the ratio $1{:}(1+t)$ being $t>0$ (other combinations have lower fitnesses). Now we want to know how haplotype *ac* can increase its frequency in the population (Fig. 1). Bodmer showed that if *AC/ac* individuals are the fittest, the frequency of *ac* will increase if both loci are closely linked. The dependence between the frequency of *ac* as a function of the number of generations (n) for different values of extra *ac/AC* fitness (t) over the prevailing double homozygote (*AC/AC*) and of the recombination fraction (γ) is given by:

$$(ac)^{(n)} = \left[\frac{(1+t)(1-\gamma)}{1}\right]^{n} (ac)^{(0)}$$

Where $(ac)^{(0)}$ corresponds to the frequency of *ac* at the starting generation. The function $(ac)^{(n)}$ increases monotonically if $t>\gamma/(1-\gamma)$. From this inequality it is also clear that the greater t and the smaller γ are, the faster the frequency of *ac* will increase. A less stringent case involving the rise of an association between an allele at a polymorphic level at one locus and a less frequent allele at another has been proposed by Kimura (1956).[21] In fact, the existence of compensatory variants, as discussed previously, opens the possibility that many "neutral" genotypes may exist at a polymorphic level in the population which in combination with rarer variants may fulfil the condition required to increase the frequency of particular advantageous haplotypes.

Close linkage directly implies low recombination but not necessarily physical proximity. However, suppose that a new gene order *ACB* appears by inversion (small inversions are very common in yeast[22]). This may hinder recombination between loci *A* and *C* by a simple mechanical interference. In the case of a chromosomal inversion that prevents recombination $t>0$ for *ac* would be enough for starting haplotype invasion. As pointed out by Crow and Kimura (1970)[23], Bodmer's condition is too stringent because *ac* is also produced by recombination of double heterozygotes in repulsion. As the frequency of *aBc* increases, a derived mutant *acB* can also spread in the population even more easily than *aBc* because linkage is now tighter, at least at the beginning of spread when the inversion mutant is still rare. In time, the frequency of *acB* increases and recombination, in individuals carrying the same gene order, becomes more likely. However, the recombination fraction γ between *a* and *c* has dropped due to their "novel" proximity. Selection can still act to counter any rearrangement breaking up such an order. In line with this, clusters of essential genes in yeast are in regions of low recombination and large clusters have low recombination rates.[24] In the end, slowly or rapidly, the neo-haplotype carrying the new order can sweep out the ancient one. The perspective outlined above is a snapshot of what is expected to be a long and stepwise process. It is conceivable that during the formation of a tight cluster, in physical terms, more than one step to reduce distance between the relevant genes will be required. Note also that a balance between the benefits of certain haplotypes and the recombination rate is required. Extremely low or absent recombination may reduce the efficacy of selection through a series of mechanisms that will not be dealt with here.[25]

Let us now explore another possibility within the context of gene duplication.[26] During duplication, the ancestral locus will carry active alleles in all members of the population and the descendant (duplicate) will be represented by a single gene in a few heterozygous individuals (according to the rate of duplication). All other individuals at this latter locus will effectively be

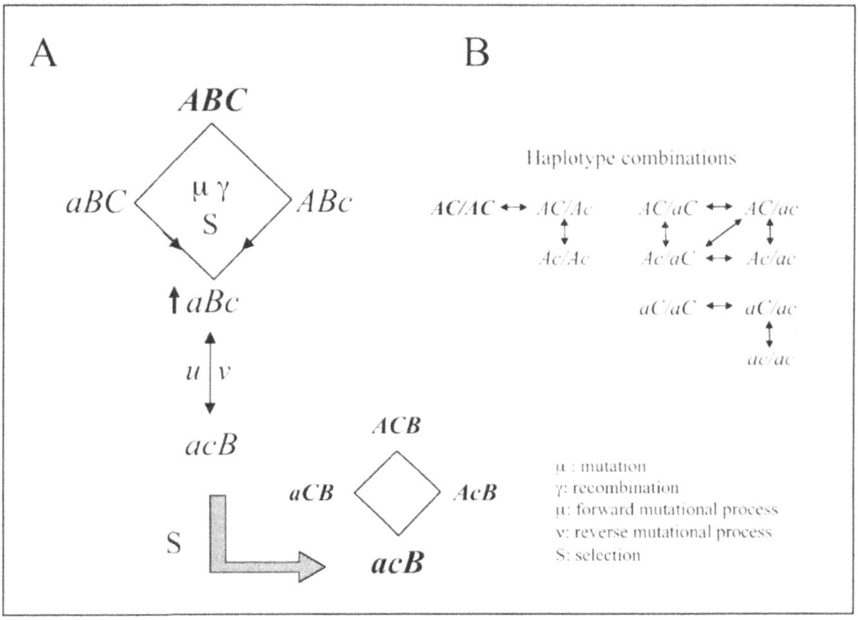

Figure 1. A pathway towards fixation of a new gene order in a population. Panel A: Let us focus on a sexual diploid population. Genes *A* and *C* encode interacting partners (one of them titrating) and are arrayed as *ABC*. *B* is an intervening sequence. Alleles *A* and *C* are frequent (in bold) while *a* and *c* are rare. The rare haplotype *ac* confers an advantage so that individuals with other haplotypes have lower fitnesses. As apparent from the reversible reactions in the lozenge, there is interconversion of haplotypes by recombination and mutation. This is better seen in the matrix represented in Panel B. Horizontal and vertical double-headed arrows link haplotype combinations that are one mutational step away one from another. The diagonal arrow represents the possibility of recombination in the double heterozygotes (coupling *AC/*ac and repulsion *Ac/aC*). The frequency of *ac* will increase if loci *a* and *c* are tightly linked. So, this is an "automatic selection for close linkage between interacting loci". Assume that a new gene order *ACB* appears by inversion and that genes *A* and *C* form a nonrecombining block because of their physical proximity. As the frequency of ***aBc*** increases a derived mutant ***acB*** will spread in the population even more easily because linkage is tighter than above. Order inversion is simply pulling out the neo-haplotype ***acB*** from the set of interconversion reactions (big lozenge). Notice that inversion reactions are also taking place at the remaining vertices of the lozenge but they are of minor importance because selection is favoring only ***aBc*** or ***acB***. In time, the frequency of ***acB*** increases and recombination becomes more likely but γ between *a* and *c* has dropped due to their proximity. Linkage desequilibrium favoring ***acB*** can be maintained by direct selection on the haplotype and/or by selecting modifiers for low (but not absent) recombination. Eventually, the neo-haplotype carrying the new order can sweep out the ancient one.

null homozygotes.[27] Thus, critical to evolution by gene duplication is an initial phase during which a duplicated copy of the gene must rise to a high enough frequency in the population to undergo mutational processes that may confer a selective advantage (subfunctionalization and neofunctionalization) that justify the retention of the duplicated copy. Subfunctionalization refers to a process in which each copy of the duplicate gene carries out a unique set of (degenerate) complementary sub-functions while neofunctionalization is connected with the evolution of new beneficial functions at the expense of the original one.[27] In many instances these duplicates are able to evolve functions that eventually lead to a replacement of the original genes. This has been prompted as an explanation of the formation of analogous enzymes, those with the same activity but structurally unrelated.[28]

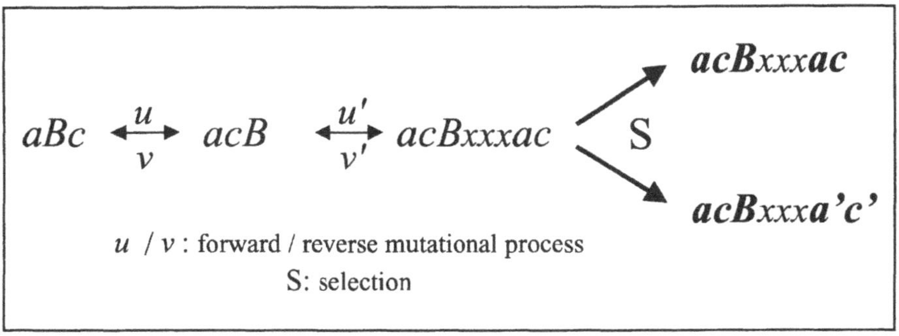

Figure 2. A "selfish DNA" pathway to the clustering of functionally related genes. As discussed in the text, selection has ways to increase the frequency of a particular haplotype or a new gene order. In our *ABC* example, suppose that a new mutant having a duplicated copy of the gene encoding the dosage sensitive component (say *c*) has a selective advantage. This is conceivable if the duplicated gene undergoes subfunctionalization or neofunctionalization. When *a* and *c* are very close in physical terms, as in *acB* but not in *aBc*, they are more likely to be duplicated together avoiding dosage imbalance. The pair *ac* would act as an "addiction module" since the gene encoding component *c* cannot duplicate alone. From this perspective, both genes *a* and *c* must remain active until the one coding for *c* breaks the "commensalist couple" (i.e., loses its dosage sensitive character, for example by losing its interaction capacity with other members of the complex).

As discussed earlier, selection has ways to increase the frequency of a particular haplotype or a new gene order. In our *aBc* example, postulate now that a new mutant having a duplicated copy of the gene encoding a dosage sensitive component (say, *c*) has a selective advantage. When *a* and *c* are very close as in *acB* they are more likely to be duplicated together avoiding dosage imbalance. In fact, in an *aBc* configuration there are also attempts of gene duplication but, when the intervening sequence *B* is long, in most cases the relevant genes *a* and *c* are expected to undergo duplication alone. In these conditions, duplication of gene *a* is neutral while duplication of *c* is deleterious. Thus, one can say that *ac* form a "commensalist couple". That is: *c* is indifferent to *a* but the former "takes advantage" of the association with the latter (i.e., during coduplication). This can be assimilated to a selfish DNA behavior.[2] The pair *ac* would act as an addiction module in the sense that gene *c*, encoding the titrating component, cannot duplicate alone. The irreversible effect of selection for a dosage increase may lead/contribute to a replacement of the original population (Fig. 2).

It is often assumed that gene duplication per se is selectively neutral and that, following a duplication, one of the gene copies is freed from purifying selection, which allows divergence. Kondrashov et al (2002)[29] have searched for systematic evidence of accelerated evolution after duplication using bacterial, archaeal and eukaryotic genomic data. They have found that most paralogs (duplicates) typically evolve at similar rates indicating that the original gene and its copy are subject to purifying selection. However, this selection is weaker than the selection affecting unduplicated orthologs that have diverged to the same extent as the analyzed paralogs. These results indicate that recently duplicated paralogs evolve faster than orthologs with the same level of divergence and similar functions, but apparently do not experience a phase of neutral evolution as commonly thought. In line with the ideas presented here, they have hypothesized that gene duplications that persist in an evolutionary scale are beneficial from the time of their origin, due primarily to a protein dosage effect in response to variable environmental conditions.[29]

It is conceivable that a mutant gene order, which is neutral at first, can also increase in frequency due to random events that a population may undergo. Theory shows that this random genetic drift is more likely when there is a reduction in the effective population size. In our context, since duplication of the newborn cluster is advantageous, the new gene order can replace the old one even if initially the new gene order increased its frequency by drift. The ultimate fate of the original genes and the copies will depend on a series of factors. Subfunctionalization will be the preferred preservation mechanism of the duplicate in small populations and the copy will tend to be linked to the original gene.[27] Neofunctionalization is more prevalent as the population size increases and the duplicates will tend to freely recombine (i.e., they will be unlinked). Multiple rounds of the processes described here and above (involving other genes) will increase cluster size in an evolutionary time-scale.The balance hypothesis[14,15] predicts that genes encoding interacting proteins should remain as single copies or both should undergo gene duplication. Indeed, when Papp et al (2003)[16] analyze pairs of genes coding for interacting subunits, they find a large excess of pairs with the same number of paralogs. This does not imply automatically that synteny of the gene pairs has to be preserved or be detectable in all cases. For instance, if a functional allele rises its frequency by chance or necessity at a duplicated locus, the ancestral copy can become a pseudogene and even disappear in time, giving the false appearance that the remaining/detected gene is "alone" (see ref. 30 and references therein).

One may argue that aneuploids rarely have a selective advantage. However, this assertion is flawed because it is the result of an ascertainment bias. Indeed, the aneuploids that we have been able to detect in the laboratory or in the hospital, at least in the pregenomic era, are those with fitness defects. We all have been at one moment or another polyploids and aneuploids, selected for and able to replace original populations. Evidence for this is compelling[31]. Before closing this section, it is interesting to discuss the effect of gene and genome duplications within the context of robustness against heterozygous and null mutations. Deletion of a gene may have a small phenotypic effect, owing to compensation. This is due either to the existence of duplicate genes or of alternative metabolic/signaling pathways. Gu et al (2003)[32] have carried out a genome-wide evaluation of the role of duplicate genes in robustness against null mutations in *Saccharomyces*. They estimate that in yeast at least a quarter of those gene deletions that have no phenotype are compensated by duplicates, supporting the notion of massive subfunctionalization in yeast. A diploid being will have four subfunctionalized copies coding for essentially the same protein, which imparts robustness in cases of heterozygous (single-gene) deletions or duplications. Thus, an increase in the number of subfunctionalized copies is expected to increase robustness in general. It is accepted that yeast underwent a round of genome duplication (ref. 32 and references therein, see below). In light of this, if we assume that the two diploids that fused were significantly different (leading to a huge advantage), one can also expect gene clustering by nonrandom deletion of genes from the paralogous chromosomes where interacting dosage sensitive pairs would be more likely retained to avoid dosage imbalances. This provides a further mechanism of nonrandom gene clustering.

The interplay between the physicochemical and genomic strategies to avoid stoichiometric imbalances may explain why the patterns of physical associations change from one organism to another.[11] From the perspective depicted here, it does not follow that a gene encoding the "titrating" component (i.e., gene *c*) must be linked to a gene coding for the same nontitrating partner (i.e., *a*) in all organisms. It might well be linked to a gene encoding another directly interacting nontitrating component with the same result.

Gene Clustering and Dosage Compensation

Changes in gene expression in aneuploids can be either direct/positive or inverse/negative. In most cases, one chromosomal dose leads to target gene up-regulation while the same segment in trisomy may lower target gene expression up to 67% of the normal diploid level (inverse effect).[33,34] In *Drosophila*, a phenotypic screen for dosage effects of the autosomes on the expression of white (an X-linked eye color gene) detected multiple regions that modulate its expression with predominant inverse effects[35] and in a few cases, there was a positive impact (see Table 1 in Chapter 7). Since the chromosomal dosage of the target gene assessed in these experiments is held constant the dosage effects are expected to result from the action of regulatory loci. As pointed out by Birchler et al (2001)[36] stoichiometric imbalances may explain why aneuploidy has a greater impact on target gene expression than changes in whole ploidy, as different ploidy levels would lead to similar relative expression of the different subunits in a complex. However, the most relevant finding of this type of study is that in many cases genes encoded on the varied segment exhibit dosage compensation. As a matter of example, when a small region including the maize *Adh1* gene is varied, a direct correlation between gene dosage and expression is found. However, an increase from one to three doses of another (normally) linked region excluding *Adh1*, produces a negative effect on ADH levels and in the whole of the chromosome arm (i.e., covariation), the two effects compensate each other.[37] Thus, covariation of both loci keep ADH levels stable. Other examples of dosage compensation of different target genes including *Glb1*, *Adh2*, *Sus1*, *Sh1* and *zein* have been described.[34] These authors also noticed that evolutionarily related genes as *Adh1* and *2* and *Sus1* and *Sh1* exhibit distinct patterns of dosages effects suggesting divergence in their epistatic interaction with linked loci. Dosage compensation is also obtained with dosage series of segments surrounding the *Adh* gene in Drosophila (Fig. 3).[38]

Last but not least, it is known that two interacting polypeptides in one organism can exist as a single polypeptide in another. The latter are called Rosetta stone proteins. A great deal of composite proteins in eukaryotes, including yeast and higher organisms, have recently been described.[39] They can be considered as examples of the maximum degree of clustering of the underlying coding sequences.

Conclusions: An Evolutionary Perspective

In an evolutionary time-scale eukaryotic genomes tend to grow through whole[31,40] or segmental duplications.[30] Recent works suggest that at a topological level protein-protein interaction networks are sparse, organized in such a way that only a few components interact with many other (hubs) while most of them are poorly connected.[41] This specific topology allows us to think of the genome as a patchwork of clusters, which would be raw duplication units, one "insulated" from each other. This modular construction is a central issue concerning evolvability which is defined as the organism's capacity to generate heritable phenotypic variation[42] or its ability to facilitate change.[43] This implies that alterations in one module (i.e., duplication) does not affect functioning of other, which minimizes interference between optimization of different modules.[44] This is a reflection of what Kirschner and Gerhart call "weak linkage" (low interdependence among processes) which reduces constraints and increases evolvability. From the perspective outlined here, just after duplication of a cluster there will be little or no stoichiometric imbalance. Later, in a classical scenario, most duplicated genes degenerate while one or a few may undergo neo or subfunctionalization eventually becoming unable to interfere stoichiometrically.

All in all, notice that selection for physical proximity between interacting loci can be driven by:

1. the advantage of a particular haplotype (supergene scenario + inversion),
2. nonrandom deletion of genes from the paralogous chromosomes after whole genome duplication (preferential retention of dosage sensitive pairs to avoid dosage imbalances) and secondarily;
3. by an advantage conferred by duplication of the cluster.

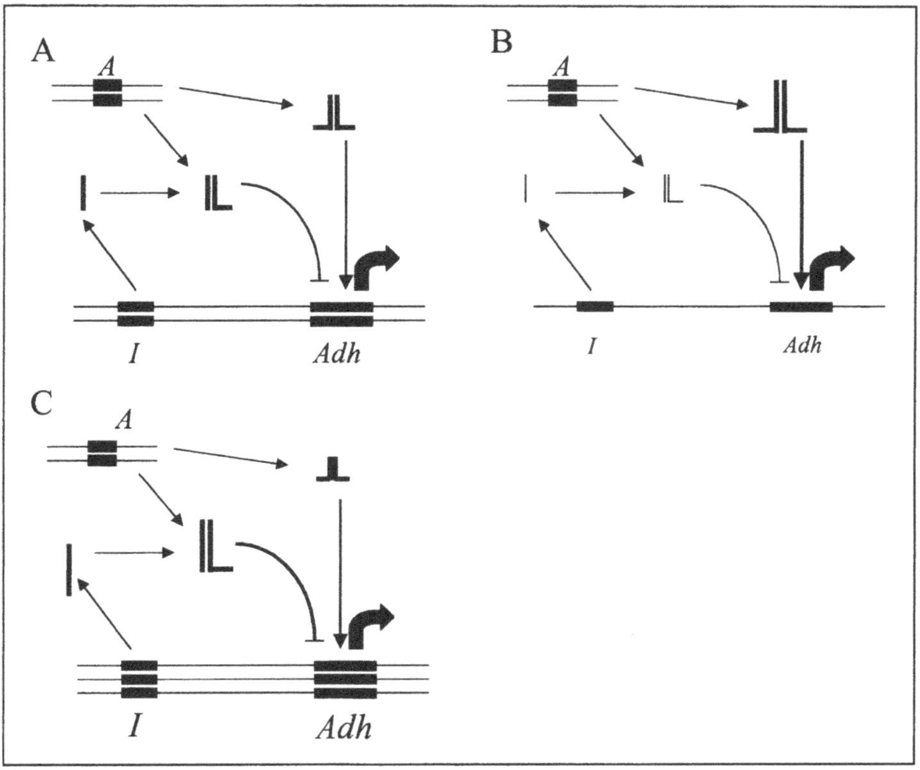

Figure 3. Gene clustering and dosage compensation: an example. Compensation has been attributed to inverse dosage effects. When there is a single copy of the region containing the target gene, linked inverse regulators of the gene will also be hemizygous and lead to up-regulation of the former. In trisomy, there will be three copies of both the structural gene and its regulator that will down regulate the former. In both cases, almost diploid levels of target gene expression are expected. The figure displays a possible molecular translation of the above. Panel A represent a diploid condition where the *Adh* gene is linked to a gene encoding the product I (inhibitor). The latter (I) interacts with the protein A encoded by a non linked locus. AA homodimers activate *Adh* expression while AI heterodimers are unable to do so. Their balance lead to the normal diploid expression. In monosomy of the *Adh-I* region (Panel B), there is a lower I concentration and a higher activator AA concentration that will double expression of the single *Adh* copy, yielding a compensated Adh level. In the trisomy of *Adh-I* there will be a higher AI concentration and less AA (Panel C). Thus, expression of the three copies of *Adh* yield the almost normal diploid level. This is clear when transcription is proportional to the signal AA/AI and to the number of copies of the target gene. Here I invoked a foreigner *A* gene. However, a similar reasoning holds for the action of (I) alone if it could inhibit transcription directly.

On the other hand, clustering, as well as duplication of a newly formed cluster, depend to a great extent on random mutational events. However, a new gene order may impart resilience to the detrimental effects that gene duplication or deletion can cause. This secondary advantage could rise a barrier against destruction of the cluster. This can be assimilated to a selection for evolvability, as outlined in Casjens (1992)[45] and remind us of the notion of biological "spandrels" prompted by Gould.[46] New gene order might be neutral during further duplication. This short-term neutrality (within the lifetime of the individual) is advantageous within the context of evolution because the newborn aneuploid is a raw material.

Acknowledgements

I wish to thank N. Le Noverre for bringing into my attention the interesting Bray/Lay's paper and J. Cocquet, S. Caburet, D. Vaiman, A. Wagner, L. D. Hurst and Sir W. Bodmer for helpful comments on the manuscript.

References

1. Dandekar T, Snel B, Huynen M et al. Conservation of gene order: A fingerprint of proteins that physically interact. Trends Bioch Sci 1998; 23:324-328.
2. Lawrence J. Selfish operons: The evolutionary impact of gene clustering in prokaryotes and eukaryotes. Curr Opin Genet Develop 1999; 9:642-648.
3. Lawrence JG. Shared strategies in gene organization among prokaryotes and eukaryotes. Cell 2002; 110:407-413.
4. Blumenthal T. Gene clusters and polycistronic transcription in eukaryotes. Bioessays 1998; 20:480-487.
5. Keene JD, Tenenbaum SA. Eukaryotic mRNPs may represent posttranscriptional operons. Mol Cell 2002; 9:1161-1167.
6. Bodmer WF, Parsons PA. Linkage and recombination in evolution. Adv Genet 1962; 11:1-100.
7. Cooper DN. Analysis of the Human Genome. Cooper DN, Farzaneh F, eds. Oxford, England: BIOS Scientific, 1995.
8. Cohen BA, Mitra RD, Hughes JD et al. A computational analysis of whole-genome expression data reveals chromosomal domains of gene expression Nat Genet 2000; 26:183-186.
9. Ge H, Liu Z, Church GM, Vidal M. Correlation between transcriptome and interactome mapping data from Saccharomyces cerevisiae. Nat Genet 2001; 29:482-486.
10. Lercher MJ, Urrutia AO, Hurst LD. Clustering of housekeeping genes provides a unified model of gene order in the human genome. Nat Genet 2002; 31:180-183.
11. Lee JM, Sonnhammer EL. Genomic gene clustering analysis of pathways in eukaryotes Genome Research 2003; 13:875-882.
12. Cremer T, Cremer C. Chromosome territories nuclear architecture and gene regulation in mammalian cells. Nat Rev Genet 2001; 2:292-301.
13. Dundr M, Hoffmann-Rohrer U, Hu Q et al. A kinetic framework for a mammalian RNA polymerase in vivo. Science 2002; 298:1623-1626.
14. Veitia RA. Exploring the etiology of haploinsufficiency. Bioessays 2002; 24:175-184.
15. Veitia RA. Nonlinear effects in macromolecular assembly and dosage sensitivity. J Theor Biol 2003; 220:19-25.
16. Papp B, Pal C, Hurst LD. Dosage sensitivity and the evolution of gene families in yeast. Nature 2003; (in press).
17. Bray D, Lay S. Computer-based analysis of the binding steps in protein complex formation. Proc Natl Acad Sci USA 1997; 94:13493-13498.
18. Janin J. Principles of protein-protein recognition from structure to thermodynamics. Biochimie 1995; 77:497-505.
19. Veitia RA. A sigmoidal transcriptional response: Cooperativity, synergy and dosage effects. Biol Rev 2003; 78:149-170.
20. Darlington CD, Mather K. The elements of genetics. London, England: Allen and Unwin, 1949.
21. Kimura M. A model of genetic system which leads to close linkage by natural selection. Evolution 1956; 10:278-287.
22. Seoighe C, Federspiel N, Jones T et al. Prevalence of small inversions in yeast gene order evolution. Proc Natl Acad Sci USA 2000; 97:14433-14437.
23. Crow JF, Kimura M. 1970. In Introduction to theoretical population genetics. Pp 204-205. New York (USA). Harper and Row Publishers.
24. Pal C, Hurst LD. Evidence for coevolution of gene order and recombination rate. Nat Genet 2003; 33:392-395.
25. Marais G. Biased gene conversion: Implications for genome and sex evolution. Trends Genet 2003; 19:330-338.

26. Ohno S. Evolution by gene duplication. George Allen and Unwin London, ed. New York, USA: Springer-Verlag, 1970.
27. Lynch M, O'Hely M, Walsh B et al. The probability of preservation of a newly arisen gene duplicate. Genetics 2001; 159:1789-1804.
28. Galperin MY, Walker DR, Koonin EV. Analogous enzymes: Independent inventions in enzyme evolution. Genom Res 1998; 8:779-90.
29. Kondrashov FA, Rogozin IB, Wolf YI et al. Selection in the evolution of gene duplications. Genome Biology 2002; 3:RESEARCH0008.
30. Gu X, Huang W. Testing the parsimony test of genome duplications: A counterexample. Genom Res 2002; 12:1-2.
31. Makalowski W. Are we polyploids? A brief history of one hypothesis. Genom Res 2001; 11:667-670.
32. Gu Z, Steinmetz LM, Gu X et al. Role of duplicate genes in genetic robustness against null mutations. Nature 2003; 421:63-6.
33. Birchler JA, Newton KJ. Modulation of protein levels in chromosomal dosage series of maize: The biochemical basis of aneuploid syndromes. Genetics 1981; 99:247-266.
34. Guo M, Birchler JA. Trans-acting dosage effects on the expression of model gene systems in maize aneuploids. Science 1994; 266:1999-2002.
35. Sabl JF, Birchler JA. Dosage dependent modifiers of white alleles in Drosophila melanogaster. Genetical Res 1993; 62:15-22.
36. Birchler JA, Bhadra U, Bhadra MP et al. Dosage-dependent gene regulation in multicellular eukaryotes: Implications for dosage compensation aneuploid syndromes and quantitative traits. Develop Biol 2001; 234:275-288.
37. Birchler JA. The genetic basis of dosage compensation of alcohol dehydrogenase-1 in maize. Genetics 1981; 97:625-637.
38. Birchler JA, Hiebert JC, Paigen K. Analysis of autosomal dosage compensation involving the Alcohol dehydrogenase locus in Drosophila melanogaster. Genetics 1990; 124:677-686.
39. Enright AJ, Ouzounis CA. Functional associations of proteins in entire genomes by means of exhaustive detection of gene fusions. Genome Biol 2001; 2:research00341-0034.7.
40. Langkjaer RB, Cliften PF, Johnston M et al. Yeast genome duplication was followed by asynchronous differentiation of duplicated genes. Nature 2003; 421:848-852.
41. Featherstone DE, Broadie K. Wrestling with pleiotropy: Genomic and topological analysis of the yeast gene expression network. Bioessays 2002; 24:267-274.
42. Kirschner M, Gerhart J. Evolvability. Proceedings of the National Academy of Sciences USA 1998; 95:8420-8427.
43. West-Eberhard MJ. Evolution in the light of developmental and cell biology, and vice versa. Proc Natl Acad Sci USA 1998; 95:8417-8419.
44. Pepper JW. The evolution of evolvability in genetic linkage patterns. Biosystems 2003; 69:115-126.
45. Casjens S, Hatfull G, Hendrix R. Evolution of the dsDNA tailed-bacteriophage genomes. Seminars Virol 1992; 3:383-397.
46. Gould SJ. The exaptive excellence of spandrels as a term and prototype. Proc Natl Acad Sci USA 1997; 94:10750-10755.

CHAPTER 9

Lessons from a Genetic Network about the Causes of Dominance

Andreas Wagner

Abstract

I review recent findings from the stoichiometric analysis of metabolic networks. These find ings show that the physiological theory of dominance, which explains dominance in metabolic genes from kinetic properties of biochemical pathways and redundant allelic functions is incomplete. Stoichiometric analysis of metabolic networks indicates that a mix of causes—local allele redundancy and global network properties—is responsible for dominance in metabolic genes.

Introduction

Dominance is a special case of a genetic system's robustness against mutations. It can be defined as robustness to a 50% change in gene dosage, and is caused by the redundant functions of two alleles at a diploid gene locus. While robustness to a 50% change in gene dosage is remarkable, it is also worth pointing out that robustness against more drastic genetic change is widespread. For example, more than one third of all synthetic null (gene knockout) mutations in the yeast *Saccharomyces cerevisiae* have weak or no phenotypic effects under standard laboratory conditions.[1,2] These are mutations where either two copies of the gene in a diploid cell or one copy in a haploid cell are eliminated. They completely eliminate a gene from the genome—a 100% change in gene dosage. One might think that weak phenotypic effects of gene knockout mutations are due to redundant gene functions, not among alleles at one locus, but among duplicate genes. However, between 40 and 77% of yeast genes with weak knockout effects are single copy genes.[3,4] This translates into hundreds to thousands of single copy genes with weak phenotypic effects. Similar numbers exist in the nematode worm *Caenorhabditis elegans*.[5] These findings do not negate the role of gene duplications and gene redundancy in mutational robustness—both are undoubtedly important.[3,6] However, they indicate that simple redundancy of parts can not be the only source of mutational robustness.

Arguably, most knockout mutations with no phenotypic effects are also examples of the wild-type gene's dominance, where a 50% reduction of gene dosage would have no phenotypic effect. And if many such cases of gene knockout mutations involve single copy genes, it follows that gene redundancy cannot be the only source of dominance. Unfortunately, the reasons why a single copy gene may have no phenotypic effect are generally poorly understood, with one exception: a growing body of work on genetic perturbations in metabolic networks. I will here review some relevant background, as well as implications of this work for our understanding of mutational robustness in general and dominance in particular.

The mechanistic reasons for dominance are best understood for metabolic genes acting in metabolic pathways, where dominance results from the cooperation of multiple enzymes in producing the output of a pathway and the flux of matter and energy through the pathway.[7] However, this framework predicts that complete elimination of a gene from a pathway—when considered in isolation—would have grave phenotypic effects, because it would block the entire pathway. In contrast, the work I review below suggests that considering pathways in the larger context of metabolic networks changes this picture. Blocking individual pathways may have little phenotypic effects, because a metabolic network as a whole may be able to reroute metabolic flux around the pathway. This suggests that an improved understanding of genetic interactions in genetic networks—whether metabolic or regulatory—can teach us much about the mechanistic causes of dominance.

Background of Stoichiometric Network Analysis

Metabolic pathways are but figments of the complex reaction networks that sustain the living. Ultimately, to understand robustness in such networks will require understanding the functioning of large metabolic networks. A quantitative theory such as metabolic control analysis would allow such an understanding.[8] However, such understanding also has a price, namely that the theory requires much quantitative information, in particular about kinetic rate constants of enzymatic reactions. Unfortunately, with the exception of a few simple model systems, such as the metabolism of human red blood cells,[9] such information is unavailable. This holds even for the best studied metabolic networks, such as that of *Escherichia coli.* Although most if not all chemical reactions catalyzed by this bacterium are known, kinetic information is available only for a small fraction of these enzymes. Thus, even if the substantial mathematical problems of applying metabolic control analysis to large reaction networks were solved, insufficient information would be available to apply the theory to large networks.

This problem raises the question: is it possible to characterize large chemical reaction networks and their robustness even though information about the reaction rates of many enzymes is unavailable? The answer is yes. It can be done by merely examining stoichiometric properties of metabolic networks, using related approaches such as flux balance analysis[10] or elementary mode analysis.[11,12] I will briefly outline the foundation of these approaches and some of the insights they provide.

Consider the following simple reaction scheme (Fig. 1), where some substrate S external to the cell is imported and/or converted into some metabolite m_1, which can then be converted through a reversible reaction (indicated by a double-headed arrow) into metabolite m_3. m_3, in turn, is a precursor to some product P, which might be a biomass component essential for cellular growth, or it might be secreted from the cell. Alternatively to its direct interconversion with m_3, m_1 can also be converted into m_3 through a chain of two irreversible chemical reactions involving a metabolite m_2.

The most important variables in a stoichiometric analysis of chemical reaction networks are the amounts of matter that flow through each of the chemical reactions indicated by the arrows in Figure 1. These are referred to as metabolic fluxes v_i, where the subscript i corresponds to the name or index of the chemical reaction, as written above the arrows in Figure 1. The changes in concentrations of the metabolites m_i, dm_i/dt are simple functions of these fluxes. For the example of Figure 1, these are

$$\frac{dm_1}{dt} = v_s - v_1 - v_2$$
$$\frac{dm_2}{dt} = v_1 + v_3 - v_p \qquad (1)$$
$$\frac{dm_3}{dt} = v_2 - v$$

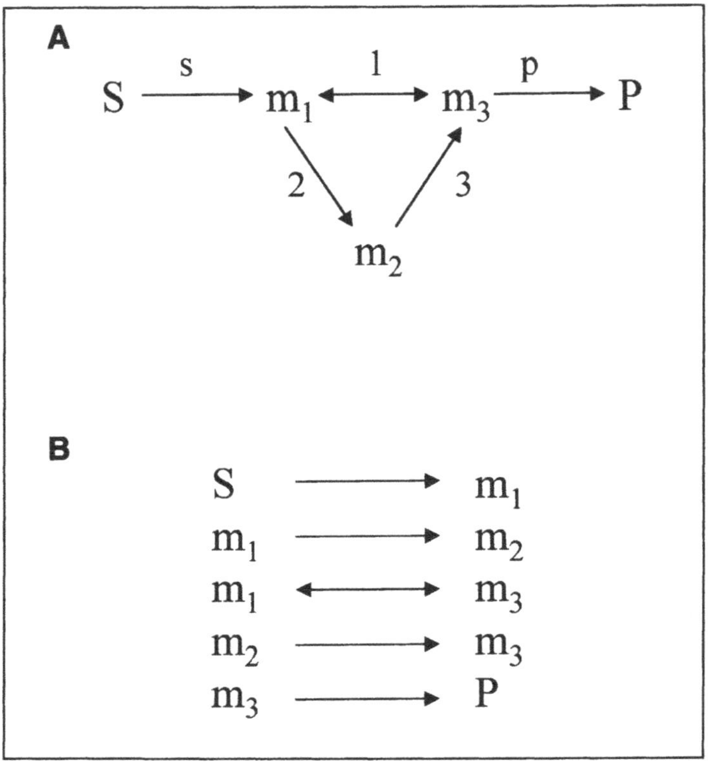

Figure 1. A simple chemical reaction scheme where an external substrate S is transformed into a metabolic product P via a series of chemical reactions involving internal metabolites m_i. The scheme is represented in two equivalent forms.

Metabolic "inputs" like S and "outputs" like P are usually referred to as "external" metabolites. The changes in their concentrations are not explicitly modeled and are distinguished from "internal" metabolites m_i, whose interconversion constitutes metabolism proper. The internal metabolites are subject to conservation of mass, which is reflected in the fact that by summing all derivatives dm_i/dt the internal fluxes v_i cancel. That is, one is left only with fluxes from and to external metabolites

$$\frac{dm_1}{dt} + \frac{dm_2}{dt} + \frac{dm_3}{dt} = v_s - v_p$$

Under constant environmental conditions, the changes in metabolite concentrations dm_i/dt must approach zero, otherwise some metabolites would disappear completely, whereas the concentration of others would approach infinity. Thus, a metabolic network approaches a steady state, in which individual metabolite concentrations do not change. This steady state is dynamic, in the sense that constant metabolite concentrations are maintained by ongoing interconversions of metabolites. Restricting the analysis only to steady states is motivated by two further observations. First, whereas transient changes in metabolite concentrations occur, for example when an environment changes, a new steady state is reached rapidly, typically within minutes.[13] Secondly, even if metabolite concentrations sometimes show more complex behavior, such as sustained oscillations, the time-averaged metabolite concentrations are constant, and thus, effectively, a steady-state is reached.

The steady state condition simplifies the mathematical treatment of a metabolic network considerably, as the following simple example shows. I begin by writing (1) in a more compact form, i.e.,

$$\frac{d\vec{m}}{dt} = S\vec{v} \qquad (2)$$

This is now a linear matrix differential equation, where $\vec{m} = (m_1, m_2, m_3)$ is the vector of internal metabolites. $\vec{v} = (v_s, v_1, v_2, v_3, v_p)$ is the vector of all fluxes, and **S** is a matrix that contains the coefficients of this differential equation, which are simply the stoichiometric coefficients of the chemical reactions for each metabolite m_i. For the simple reaction scheme of Figure 1, this matrix has the following structure.

$$S = \begin{pmatrix} -1 & -1 & -1 & 0 & 0 \\ 0 & 1 & 0 & 1 & -1 \\ 0 & 0 & 1 & -1 & 0 \end{pmatrix} \qquad (3)$$

Its columns correspond to the reactions s, 1,2,3, and p of Figure 1. Its rows correspond to metabolites m_1, m_2, and m_3. Positive and negative signs of the respective entries indicate whether a metabolite occurs in the left or right side of a chemical reaction in Figure 1B. For example $S_{32} = 1$ because metabolite m_3 is a product (occurs on the right-hand side) of reaction 2. In the notation of (2), the steady state condition (no changes in internal metabolite concentrations) can be written as

$$\frac{d\vec{m}}{dt} = 0$$

which is equivalent to

$$S\vec{v} = 0 \qquad (4)$$

For our example, this condition is equivalent to

$$\begin{aligned} v_s - v_1 - v_2 &= 0 \\ v_1 + v_3 - v_p &= 0 \\ v_2 - v_3 &= 0 \end{aligned} \qquad (5)$$

As a mere matter of convention, the fluxes corresponding to external metabolites are sometimes written on the right-hand side of (5), such that one obtains

$$\begin{aligned} v_1 - v_2 &= -v_s \\ v_1 + v_3 &= v_p \\ v_2 - v_3 &= 0 \end{aligned} \qquad (6)$$

This can be written again in matrix form

$$S'\vec{v}' = \vec{b} \qquad (7)$$

where the stoichiometry matrix S' is

$$S' = \begin{pmatrix} -1 & 0 & 1 \\ 1 & 0 & 0 \\ 0 & 1 & -1 \end{pmatrix}$$

and the vector $\vec{v}=(v_1,v_2,v_3)$, and $\vec{b}=(-v_s,v_p,0)$. The equivalence of (5) and (6) shows that the two matrix formulations also have equivalent solutions. However, there is a conceptual difference. In (4), all fluxes, including those from and to the external metabolites, v_s and v_p, are treated as variables, and the steady state flux distribution is obtained by solving for $\vec{v}$. In (7), these external fluxes are absorbed into the vector $\vec{b}$, and are thus treated as constants. (One can think of them as availabilities of a substrate in the environment, or as export rates of a product.) One thus solves only for the three internal fluxes.

I note that the above stoichoimetry matrices S and S' are unusual in several respects. First, and trivially, they contain many fewer reactions than are encountered in any network of realistic complexity. Second, the reaction scheme of Figure 1 contains only monomolecular reaction, i.e., reactions where each reaction has only one educt and only one product. Much more frequent are bimolecular reactions. Such reactions are easily incorporated into a stoichiometry matrix S. Each column of S (corresponding to one reaction) can carry as many negative entries as the reaction has educts, and as many positive entries as it has products. Third, S in this example contains only nonzero entries (+1) and (-1), but many reactions do not convert molecules in equimolar proportions. This complication is also easily incorporated. For example, if reaction 3 needed two molecules of m_3 to produce one molecule of m_2, then S_{33} would be equal to -2 instead of -1.

In addition to the flux-balance condition (4) in the steady state, any metabolic reaction network has to fulfill several additional constraints. First, fluxes can not become arbitrarily small or large, such that they need to be bounded between some real values. The reason is that only a limited amount of any one enzyme can be present, and that enzymes catalyze reactions at finite rates. Second, some reactions are irreversible and can proceed in only one direction. By convention, the respective flux v_i can not be negative. The same principles hold for the import of an external metabolite, such as a carbon source, or for the export of a metabolic end product. The respective flux has a maximal rate, which reflects factors such as the concentration of a metabolite in the environment and the transport mechanism of the metabolite.

Main Tasks of Stoichiometric Network Analysis

Stoichiometric analysis has two main tasks. First, it identifies the flux-vectors $\vec{v}$ that fulfill all the constraints on a metabolic network. These include the steady-state condition (4), as well as the additional constraints on flux magnitudes and signs just listed. Flux vectors that meet these conditions are 'allowed' fluxes, that is, fluxes that a cell can realize. Importantly there is almost never just one unique allowed flux vector. The reason is that in most metabolic reaction networks with n internal metabolites and m chemical reactions, there are many more chemical reactions than metabolites (m>>n), or many fewer equations than variables, such that (4) is massively underdetermined. If no constraints other than (4) were operational, the set of allowable fluxes would be an (n-m)-dimensional (vector) space, which is also called the null space of the stoichiometry matrix S. However, because fluxes are bounded, the allowable flux vectors occupy a bounded region in this space, as indicated in the three-dimensional caricature of Figure 2.

The second task of stoichiometric analysis is to identify regions within the set of allowable fluxes that maximize a desirable property. One example of such a property is cell growth. Consider a genetically heterogeneous population of single-celled organism that actively grow

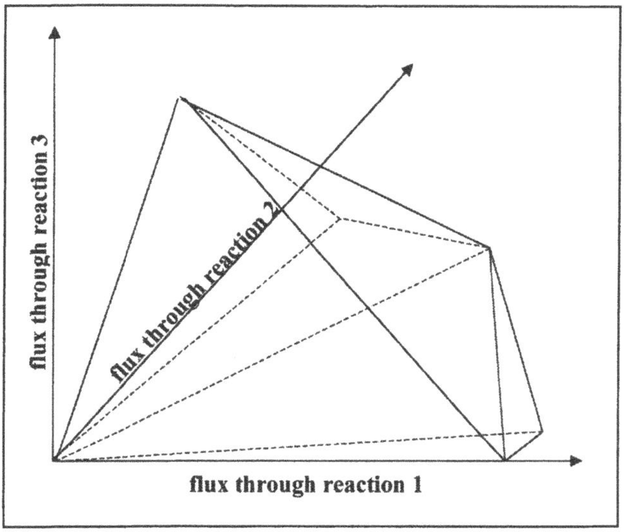

Figure 2. A schematic representation of allowed steady state fluxes for a hypothetical set of three chemical reactions (not shown). The figure illustrates that the allowed fluxes do not form a vector space but instead a bounded subset (flux cone) of a vector space, the null space of a stoichiometric matrix S.

and divide. Each cell or genotype in this population may occupy a different position in the region of allowed metabolic fluxes, because its enzymes and their expression levels under any particular environmental condition differ from those of other cells. Those cells in the population that grow at a maximal rate will outgrow all other cells, and thus come to dominate the population. For such maximal cell growth, biosynthetic precursors such as amino acids need to be made in well-defined ratios. Similary, high-energy phosphate bonds (ATP and related molecules) and redox potential (NADH and related molecules) need to be produced in balanced amounts. This means that the metabolic fluxes generating them must have particular values, which can be identified if the optimal proportions of biosynthetic precursors, energy carriers, and redox potential are known. In some well studied organisms, such as the bacterium *Escherichia coli*, these proportions are known from the biomass composition of the organism.[13] In trying to find the flux vectors $\vec{v}$ that yield maximal growth under any one environmental condition, one tries to identify one (or more) points in the realizable region of the space of fluxes (Fig. 2) that maximize or minimize some function Z of the flux $\vec{v}$. In practice, linear functions of $\vec{v}$ are most important, such that

$$Z(\vec{v}) = \sum_{i=1}^{m} c_i v_i \tag{8}$$

and one tries to find values of $\vec{v}$ that maximize Z. (There may be many flux vectors $\vec{v}$ that give the same maximal value of Z). As an aside, finding fluxes that ensure maximal growth is by far not the only application of this approach. Others include identification of fluxes that minimize ATP production—corresponding to energy-efficient growth—, or identification of fluxes that produce maximal amounts of an industrially important metabolite, such as an antibiotic. The approach is thus of great relevance for metabolic engineering of organisms in industrial biotechnology. In practice, the function Z is maximized by standard numerical techniques such as linear programming.

Note that even though a particular point in realizable flux space may be identified as optimal, it is by no means assured that a cell can achieve the desired metabolic fluxes. For example, when faced with certain carbon sources, a cell may not be able to express the required enzymes in the amounts that assure optimal growth. A case in point is the *E. coli* strain MG1655, which grows on glycerol as sole carbon source at a rate lower than predicted as optimal by this approach. However, within 40 days or 700 generations of evolution in the laboratory, the strain increases growth substantially, and the fluxes approach those predicted by theory as optimal.[15] This shows not only the power of this approach to predict optimal flux distributions, but also the power of evolution to achieve the appropriate patterns of gene regulation within short amounts of time.

To summarize, stoichiometric analysis of large enzymatic reaction networks can identify realizable metabolic fluxes, fluxes that fulfill certain boundary conditions. Within this set of all realizable fluxes, it can identify fluxes that endow a cell with properties of interest. A condition of particular relevance for evolutionary studies is that of maximal cell growth, because it is a correlate of fitness under conditions where cells actively divide.

Applications to Robustness

The approach outlined above is suitable to analyze robustness of cell growth to changes in individual fluxes. That is, under conditions where cells grow maximally, one can ask: what are the effects of reducing only one individual flux drastically, e.g., by forcing it to assume a value of zero. Biologically, such a change would correspond to a loss-of-function mutation in an individual enzyme catalyzing a particular reaction, or in a loss of its expression. Because the approach is computational, it is easily possible to determine robustness to changing—one by one—all fluxes in a network, and one can do the same for all pairwise combinations of fluxes.

In a series of papers, Edwards and Palsson[16-18] analyzed the robustness of the well-characterized chemical reaction networks in two prokaryotic organisms, *Escherichia coli* and *Haemophilus influenzae*. For *E. coli*, they assembled a reaction network comprising 436 metabolites and 736 reactions from the biochemical literature, genome sequence information, and metabolic databases.[18] They determined the allowable steady-state fluxes under the constraints listed above. Within this allowable region of fluxes, they determined those fluxes for which growth on a minimal medium under aerobic conditions with glucose as sole carbon source was maximal. Growth is maximal for fluxes that produce the necessary metabolites in proportions that correspond to the (empirically known) biomass composition of *E. coli*. (This maximal growth flux distribution can be thought of as a single chemical reaction that converts biosynthetic precursors into biomass with the least possible wastage.) With this optimal flux distribution $\vec{v}$ in hand, individual fluxes can be forced to a value of zero (corresponding to deletion of the respective enzyme-coding genes) and the resulting effect on growth can be studied. The parts of metabolism Edwards and Palsson analyzed in this way comprise 48 chemical reactions and include all of glycolysis, the pentose phosphate pathway, the tricarboxylic acid cycle, and respiration. Perhaps surprisingly, only seven of the 48 eliminated reactions turned out to be lethal. That is, they generated fluxes not allowable in steady state, or fluxes from which an essential biochemical precursor, such as an amino acid, is not produced. Two of the esssential reactions, glyceraldehyde 3-phosphate dehydrogenase and phosphoglycerate kinase, are part of the three-carbon stage of glycolysis. Three others are the first reactions of the tricarboxylic acid cycle, namely citrate synthase, aconitase, and isocitrate dehydrogenase. The remaining two essential reactions are part of the pentose phosphate shunt. These reactions are ribose-5-phosphate isomerase and transketolase. Figure 3 shows the essential reactions as part of an overview of central metabolism.

Of the remaining 41 nonessential reactions, 32 reduced growth by less than 5%, and only 9 reduced growth by more than 5%. The authors compared their results for a variety of carbon sources to experimental data that had determined the effect of deletions in enzyme-coding genes of *E. coli.* They found that the experimental data and computational predictions on growth/no-growth coincided in 86% of the cases. Thus, the computational results show good agreement with experiment.

The surprising aspect of these results is the high fraction—about two-thirds—of complete loss-of-function mutations that reduce growth by less than 5%. It bears emphasizing that there is no redundancy in this system, in the sense that there are no two enzymes that can carry out the same chemical reactions. The explanation of these findings is that flux can be shuttled through parts of a metabolic network that are not affected by blocking a reaction. Here is an example.[16]

The pentose phosphate shunt diverts metabolites from glycolysis and serves two main purposes.[19] Its oxidative branch generates NADPH for biosyntheses. (This branch can only be entered via glucose-6-phosphate.) Its nonoxidative branch generates biosynthetic precursors such as ribose-5-phosphate. It can be entered from the oxidative branch, but also from other glycolytic intermediates such as fructose 6-phosphate. Edwards and Palsson[16] blocked the first reaction of the pentose phosphate shunt (reaction zwf in Fig. 3), glucose 6-phosphate dehydrogenase, which converts glucose 6-phosphate (G6P) into D-6-phosphate-glucone-lactone (6PGA). This reaction leads into the oxidative branch of the pentose phosphate shunt. Deletion of this reaction completely blocks the oxidative branch but affects metabolic output only minimally: wild-type growth is reduced by only one percent under aerobic conditions in glucose minimal medium. However, the deletion has profound systemic consequences on the flux in this network. Before the deletion (i.e., in the wild-type state), about two-thirds of the NADPH needed is produced by the pentose phosphate shunt. Most of the NADH is produced by the tricarboxylic acid cycle. (High energy phosphate bonds are largely generated via oxidative phosphorylation.) One of the major systemic reorganizations of metabolic flow after the mutation regards NADPH production. To compensate for the blocked oxidative branch of the pentose phosphate shunt, most of the NADPH is now produced through an increased flux in the tricarboxylic acid cycle, which generates NADH. This NADH is then transformed into NADPH via a massively increased flux through the transhydrogenase reaction. The nonoxidative branch of the pentose phosphate shunt can still be entered through other metabolites of glycolysis, and thus still serves to produce sufficient quantities of biosynthetic precursors. Figure 4 contains an overview of the flux in this mutant.

A further striking result of this analysis is that the network is robust even to substantial manipulations of flux through some of the seven essential reactions.[17] That is, although complete elimination of flux through the essential reactions is lethal, substantial quantitative reductions in flux may be neutral. A case in point is the essential transketolase reaction in the pentose phosphate shunt. As long as a mutation preserves more than 15% of the wild-type flux through this reaction, growth is greater than 99.2% of the wild-type growth rate. This absence of a phenotypic effect, however, camouflages profound systemic changes that have to take place to compensate for the reduction in flux. One of these changes is again an increased production of NADPH through the tricarboxylic acid cycle and through transhydrogenase. Another change is an increased flux through glycolylytic reactions such as pyruvate kinase that absorb the reduced flux through the pentose phosphate shunt. As flux decreases below 15%, the reduced flux through transketolase limits the production of erythrose 4-phophate, an essential precursor of aromatic amino acids. The result is a reduced growth rate.

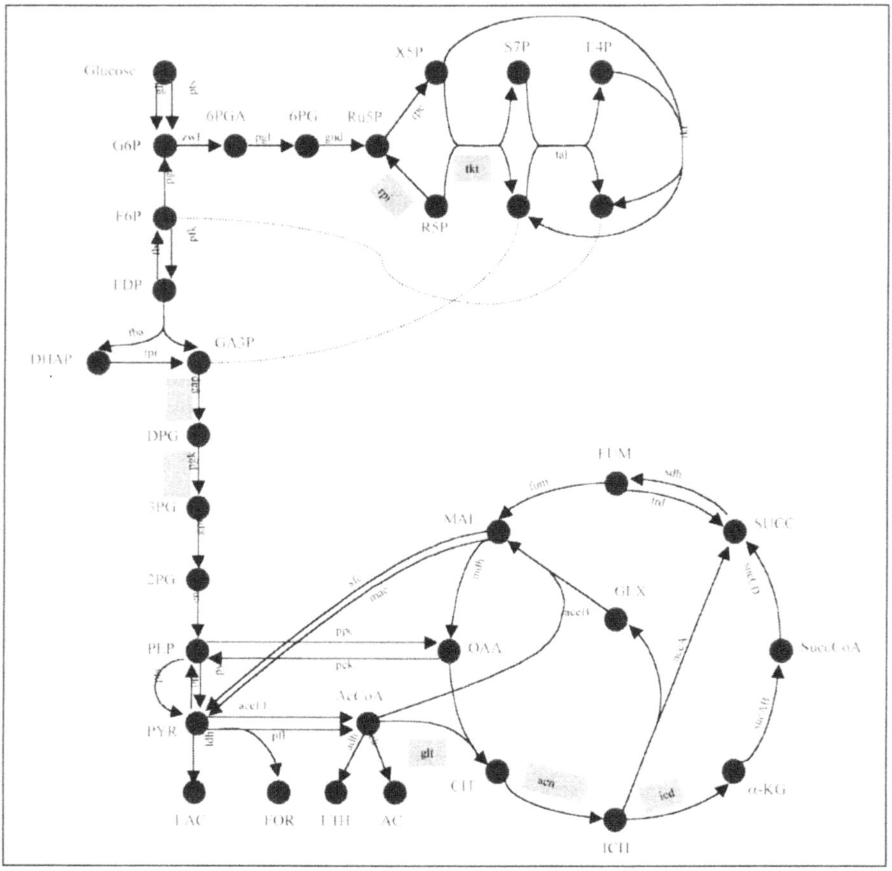

Figure 3. Central metabolic pathway reactions in *Escherichia coli*. Reactions shaded in grey are essential, that is, their removal eliminates growth according to stoichiometric analysis.[17,18]

Reactions: aceA, isocitrate lyase; aceB, malate synthase; aceEF, pyruvate dehydrogenase; ack, acetate kinase; acn, aconitase; adh, acetaldehyde dehydrogenase; eno, enolase; fba, fructose-1,6-bisphosphatate aldolase; fbp, fructose-1,6-bisphosphatase; frd, fumurate reductase; fum, fumarase; gap, glyceraldehyde-3-phosphate dehydrogenase; glk, glucokinase; glt, citrate synthase; gnd, 6-phosphogluconate dehydrogenase; gpm, phosphoglycerate mutase; icd, isocitrate dehydrogenase; ldh, lactate dehydrogenase; mae, malic enzyme; mdh, malate dehydrogenase; pck, phosphoenolpyruvate carboxykinase; pfk, phosphofructokinase; pfl, pyruvate formate lyase; pgi, phosphoglucose isomerase; pgk, phosphoglycerate kinase; pgl, 6-phosphogluconolactonase; ppc, phosphoenolpyruvate carboxylase; pps, phosphoenolpyruvate synthase; pts, phosphotransferase system; pyk, pyruvate kinase; rpe, ribulose phosphate 3-epimerase; rpi, ribose-5-phosphate isomerase; sdh, succinate dehydrogenase; sfc, malic enzyme; sucAB, 2-ketoglutarate dehyrogenase; sucCD, succinyl-CoA synthetase; tal, transaldolase; tkt, transketolase; tpi, triosphosphate isomerase; zwf, glucose 6-phosphate-1-dehydrogenase. Metabolites: 2PG, 2-phosphoglycerate; 3PG, 3-phosphoglycerate; 6PG, D-6-phosphate-gluconate; 6PGA, D-6-phosphate-glucono-‰-lactone; AC, acetate; AcCoA, Acetyl-CoA; R-KG, R -ketoglutarate; CIT, citrate; DHAP, dihydroxyacetone phosphate; DPG, 1,3-bis-phosphoglycerate; E4P, erythrose 4-phosphate; ETH, ethanol; F6P, fructose 6-phosphate; FDP, fructose 1,6-diphosphate; FOR, formate; FUM, fumarate; G6P, glucose 6-phosphate; GA3P, glyceraldehydes 3-phosphate; ICIT, isocitrate; LAC, lactate; MAL, malate; PEP, phosphoenolpyruvate; PYR, pyruvate; R5P, ribose 5-phosphate; Ru5P, ribulose 5-phosphate; S7P, sedo-heptulose; SUCC, succinate; SuccCoA, succinyl CoA; X5P, dihydroxyacetone phosphate. Figure courtesy of Jeremy Edwards. Modified with permission from Edwards JS, Palsson BO. Biotechnology Progress 2000; 16(6):927-939.

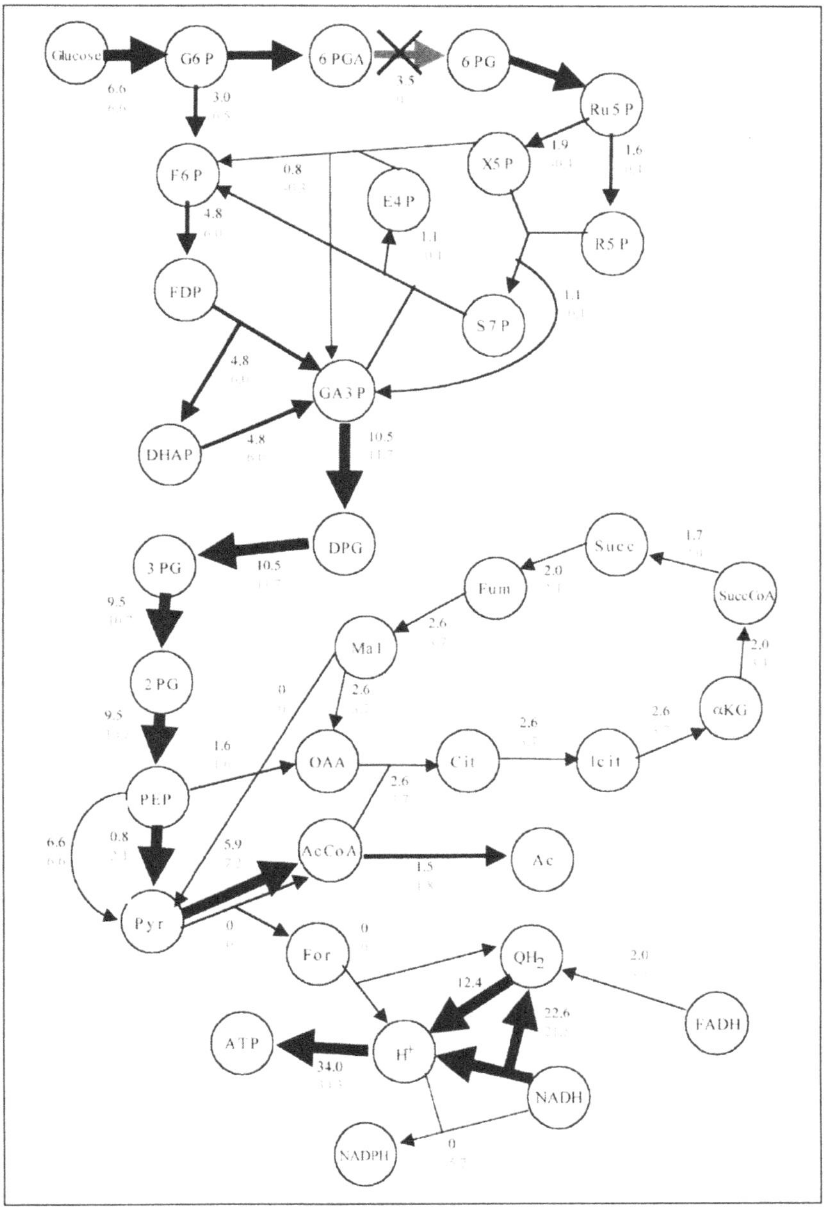

Figure 4. Rerouting of metabolic fluxes in a metabolic mutant. Growth-maximizing flux for a wild type metabolic genotype and in the *zwf-* mutant. This mutation (indicated by the crossed arrow) eliminates the glucose 6-phosphate dehydrogenase reaction that leads into the pentose phosphate shunt. Biomass yield in this mutant is only one percent lower than that in the wild-type. Thickness of arrows is proportional to the flux through a reaction in the wild-type. The upper (lower) number next to each reaction indicates the metabolic flux in the wild-type (mutant) [substrate converted h^{-1} gram dry weight $(DW)^{-1}$] at a glucose uptake rate of 6.6 mmol glucose h^{-1} g DW^{-1} and an oxygen uptake rate of 12.4 mmol oxygen h^{-1} g DW^{-1}. See legend to Figure 3 for an explanation of acronyms used. Figure courtesy of Jeremy Edwards. Modified with permission from Edwards JS, Palsson BO. Proc Natl Acad Sci USA 2000; 97:5528-5533.

The first three reactions of the tricarboxylic acid cycle, equally essential, also are quite robust to large quantitative changes in flux. For example, unless flux through the citrate synthase reaction falls below 18% of the wildtype, wild-type growth is essentially unchanged. However, as flux through this reaction is gradually reduced from the optimal wildtype level, a variety of systemic changes occur that allow wild-type growth to be sustained. They include increased flux through the pentose phosphate shunt, reduction and eventual complete elimination of flux through the pyruvate kinase reaction, (which is upstream of the blocked reaction), as well as reduction of cyclic flux through the tricarboxylic acid cycle. The cycle eventually ceases to function cyclically, and serves only to generate biosynthetic precursors. If the flux through citrate synthase falls below 18% of wild-type levels, the metabolic network can not produce sufficient α-ketoglutarate, an essential precursor of amino acids, to ensure maximal growth. This is the reason for the reduction in growth rate under these conditions.[17]

The two essential glycolytic reactions are the most restricted of the seven essential reactions. Their fluxes can not be reduced to less than 70% of the wild-type level without affecting growth substantially. Finally, it is worth mentioning that even the essential reactions can be quite insensitive to increases in flux. (I have not emphasized robustness to such "gain-of-function" mutations, because most mutations are likely to reduce flux, as they reduce either enzyme activity or enzyme expression.) For example, flux through the citrate synthase reaction can increase to 160% of wild-type without affecting growth.[17]

Edwards and Palsson also studied the metabolic reaction network of the bacterium *Haemophilus influenzae*. Their results add additional facets to the studies in *E. coli*. They used physiologial and genome information to construct a map of 488 metabolic reactions and 343 metabolites, and examined robustness of fluxes in this network to deletions in 36 central reactions. They found a larger fraction of essential genes than in *E. coli* (33% vs 14% in *E. coli*), and a smaller fraction of genes with no effects on growth when eliminated (42% vs 69% in *E. coli*). However, the networks are not straightforward to compare, because of their different sizes and features. (For example, glutamate was an essential amino acid for the *H. influenzae* network, whereas it can be produced by the *E. coli* network.)

Two aspects of the *H. influenzae* study add substantially to the *E.coli* work. First, the authors also examined the effects of multiple (double and triple) deletions of enzymatic reactions on growth. They found only 7 lethal double-mutants among 361 double mutations whose singly mutant constituents were not lethal. Similarly, among 5270 triple mutations, only 7 were lethal. That is, in large metabolic networks under defined conditions, it is quite feasible to eliminate multiple network components without destroying network function.

The second important aspect of this study is that the authors did not only analyze robustness in one environment, but they extended their analysis to several environments. The *H. influenzae* results cited above make specific assumptions about the availability and uptake rate of glutamate, an essential amino acid, and fructose, the sole carbon source. When both the availability of fructose and glutamate were varied over a range of values, the number of reactions that did not affect growth when deleted shrank from 14 to 9. When in addition the availability of oxygen was also varied, this number was further reduced to five. In other words, chemical reactions that do not affect growth in one environment may well do so in another environment.

A note of caution to all these results must be reiterated at this point. While the stoichiometric analysis reveals optimality criteria for cell growth, it does not guarantee that cells can attain the required fluxes. For example, a cell may not be able to express enzymes in the quantities necessary to ensure maximal growth. This holds in particular for unusual environmental or genetic conditions which a population has not encountered in its evolutionary history, and where no evolutionary pressure has forced an adaptive response. A possible case in point is the *E.coli* double mutant in the enzymes glucose-6-phosphate dehydrogenase and transhydrogenase.

Its growth rate is predicted to be 92% of the wild-type.[18] However, experimental results from a strain in which both genes encoding the respective enzymes were deleted show that the mutant grows only at 57% of the wild-type rate. Such quantitative discrepancies may well be due to the fact that the *E. coli* strain may never have undergone adaptive evolution in the double mutant condition.

Summary and Conclusion

The results summarized above indicate that gene redundancy is by far not the only source of robustness in metabolic networks. For instance, among 48 emulated gene knockout mutations in central metabolism of *E. coli*, only 7 revealed essential reactions. Among these essential reactions, only two can not tolerate a flux reduction of more than 50% without affecting cell growth. In other words, the vast majority of loss-of-function mutations in metabolic networks may be recessive for reasons that have nothing to do with gene redundancy. Rather, such recessivity is a property of a network capable of reallocating metabolic flux to different pathways. The canonical explanation of dominance in metabolic pathways, which emerges from metabolic control theory, is only a part of the mechanistic explanation for dominance. Its emphasis on simple parts redundancy in individual pathways renders it incomplete. It will be instructive to see whether a similar mix of causes—local gene redundancy and global network properties—accounts for robustness in other genetic networks that have come under increased scrutiny with the availability of functional genomic techniques.

Acknowledgments

I would like to thank the NIH for its support through grant GM063882-01 and the Santa Fe Institute for its continued support.

This chapter is to appear in similar form in the monograph, *Robustness and Evolvability in Living Systems*, by the author, which will be published by Princeton University Press in 2005. It is reprinted with permission from Princeton University Press.

Reference

1. Smith V, Chou KN, Lashkari D et al. Functional analysis of the genes of yeast chromosome V by genetic footprinting. Science 1996; 274:2069-2074.
2. Giaever G, Chu A, Ni L et al. Functional profiling of the Saccharomyces cerevisiae genome. Nature 2002; 418:387-391.
3. Gu Z, Steinmetz L, Gu X et al. Role of duplicate genes in genetic robustness against null mutations. Nature 2003; 421:63-66.
4. Wagner A. Mutational robustness in genetic networks of yeast. Nature Genetics 2000; 24:355-361.
5. Conant GC, Wagner A. Duplicate genes and robustness to transient gene knock-downs in Caenorhabditis elegans. Proc R Soc Lond B Biol Sci 2004; 271(1534):89-96.
6. Conant GC, Wagner A. GenomeHistory: A software tool and its applications to fully sequenced genomes. Nucleic Acids Research 2002; 30,1-10.
7. Kacser H, Burns JA. The molecular basis of dominance. Genetics 1981; 97:639-666.
8. Fell D. Understanding the control of metabolism. Miami: Portland Press, 1997.
9. Palsson BO, Lee I-D. A comprehensive model of human erythrocyte metabolism: Extensions to include pH effects. Biomed Biochim Acta 1991; 49:771-789.
10. Edwards JS, Ramakrishna R, Schilling C et al. Metabolic flux balance analysis. In: Lee S, Papoutsakis E, eds. Metabolic engineering New York: Marcel Dekker, 1999:13-57.
11. Schuster S, Hilgetag C, Woods J et al. Elementary modes of functioning in biochemical networks. In: Cuthbertson R, Holcombe M, Paton R, eds. Computation in cellular and molecular biological systems. Singapore: World Scientific, 1996:151-164.
12. Schuster S, Dandekar T, Fell DA. Detection of elementary flux modes in biochemical networks: A promising tool for pathway analysis and metabolic engineering. Trends Biotechnol 1999; 17:53-60.

13. Palsson BO, Lightfoot EN. Mathematical-Modeling of Dynamics and Control in Metabolic Networks .1. On Michaelis-Menten Kinetics. J Theor Biol 1984; 111:273-302.
14. Neidhardt FC, Umbarger HE. chemical composition of Escherichia coli. In: Neidhardt FC, ed. Escherichia coli and Salmonella: Cellular and molecular biology. Vol 1. Washington, DC: American Society of Microbiology Press, 1996:13-16.
15. Ibarra RU, Edwards JS, Palsson BO. Escherichia coli K-12 undergoes adaptive evolution to achieve in silico predicted optimal growth. Nature 2002; 420:186-189.
16. Edwards JS, Palsson BO. Systems properties of the Haemophilus influenzae Rd metabolic genotype. J Biol Chem 1999; 274:17410-17416.
17. Edwards JS, Palsson BO. Robustness analysis of the Escherichia coli metabolic network. Biotechnology Progress 2000; 16(6):927-939.
18. Edwards JS, Palsson BO. The Escherichia coli MG1655 in silico metabolic genotype: Its definition, characteristics, and capabilities. Proc Natl Acad Sci USA 2000; 97:5528-5533.
19. Stryer L. Biochemistry. 4th ed. New York: Freeman, 1995.

Chapter 10

From Beanbag Genetics to Feedback Genetics:

Bridging the Gap between Regulatory Biology and Quantitative Genetics Theory

Stig W. Omholt

Abstract

This chapter addresses the abyss that currently exists between quantitative genetics theory and regulatory biology. My claim is that despite the apparent success of quantitative genetics within evolutionary biology and production biology, the conceptual and methodological foundation of the theory as such needs a major revision if we are to hope for development of a genetics theory that is capable of actually linking genes, phenotypes and population level genetic phenomena in a causal explanatory structure. I predict that nonlinear system dynamics will make up a major part of the core of the mathematical foundation of such a future theory. The simple reason for this is that statistics is not an adequate language to describe and analyse how emergent dynamic phenomena (i.e., phenotypes) are generated by the interactions of lower-level systemic entities. As a proof of principle of the plausibility of the above considerations I analyse the phenomenon of genetic dominance. By acknowledging that genes interact in feedback structures, and making use of an appropriate mathematical language to describe and analyse such genetic structures, a whole new research field within the disciplinary frames of quantitative genetics opens up.

Characteristics of Current Quantitative Genetic Theory

Conceptual and Methodological Foundation

Quantitative genetics holds a prominent position in such fields as plant and animal breeding, evolutionary biology, medicine and psychology.[1-3] The discipline addresses mainly physiological, morphological and behavioural phenotypic metric traits showing continuous variation. Classical quantitative genetics considers these traits to be controlled by a very large (infinite) number of genes, each of small effect.[4] At the intra-locus level the genotypic effects of the alleles of a diallelic locus are defined to be additive or dominant, depending on whether the genotypic value of the heterozygote is equal to the mean of the two homozygotes or not (Fig. 1a). The genotypic effects of the loci contributing to a metric character show inter-locus additivity when their individual effects can be summed to give the genotypic value of the individual. When this is not the case there are one or more inter-locus interactions causing epistatic genetic effects in addition to the additive ones. Environment is normally assumed to be additive to the genotype.[1,5]

The Biology of Genetic Dominance, edited by Reiner A. Veitia. ©2006 Eurekah.com.

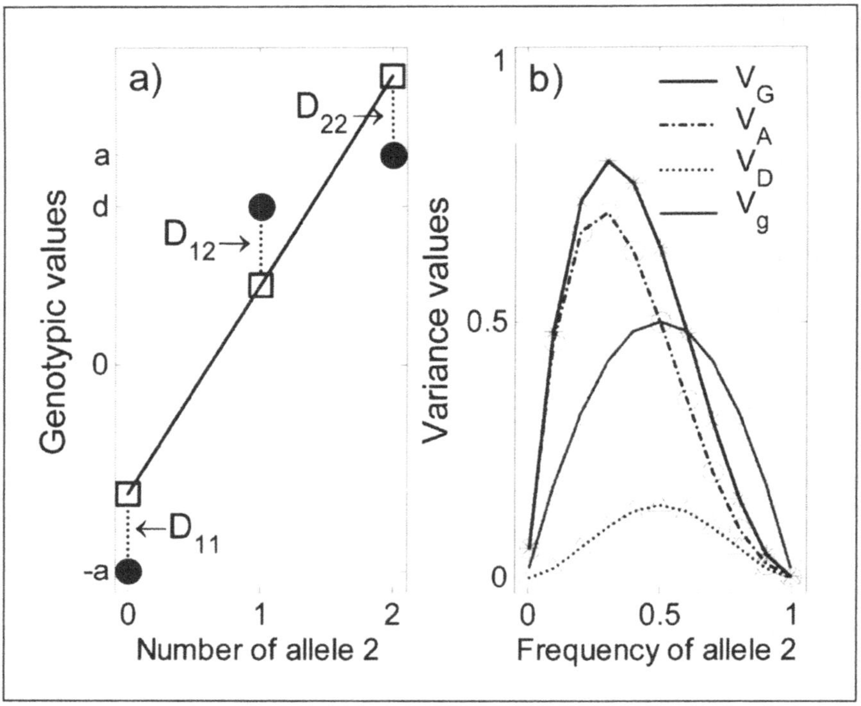

Figure 1. Basic diallelic one-locus quantitative genetics under Hardy-Weinberg conditions. a) The relationship between the genotypic values G_{11}=-*a*, G_{12}=*d*, and G_{22}= *a* with frequencies, q^2,$2pq$, and p^2, and number of dominant alleles (allele 2). *d* is the genotypic value of the heterozygote and measures the dominance deviation in the physiological sense. $D_{ij}=G_{ij}-\hat{G}_{ij}$ is the dominance deviation in the statistical sense. It measures the difference (i.e., error) between actual (G_{ij}=●) and predicted ($\hat{G}_{ij}$=□) genotypic values based on a least-squares linear regression of genotypic value on gene content equation $\hat{G}_{ij} = \hat{\alpha} + \hat{\beta} N_{ij}$ is the genotype's number of dominant alleles, $\hat{\beta} = \frac{\sigma(N,G)}{\sigma_N^2}$ is the regression coefficient, $\hat{\alpha} = \overline{G} - \hat{\beta}\overline{N}$ is the intercept value, $\overline{G}=p^2a + 2pqd - q^2a$ is the genotypic mean, and $\overline{N} = p^2 \cdot 2 + 2pq \cdot 1 + q^2 \cdot 0$ is the mean proportion of dominant alleles (a = 1, d = 0.75, p = q = 0.5). b) Variance components that follow from the presumed linear relationship between genotypic value and gene content depicted in the left panel (over the whole allele frequency range). V_G = total genotypic variation, V_A = additive variance expressing how much of the variation that can be explained by the linear assumption, and V_D = the variance of the dominance deviations (D_{ij}), which in this case is equal to the error variance (V_G = V_A + V_D). The heritability (h^2) is normally expressed as, $h^2 = V_A / V_p$ where V_p is the phenotypic variance. In this example there is no influence from the environment, so $h^2 = V_A / V_G$. V_g shows the total genotypic variance when there is no dominance (d = 0) with all other parameters equal. These relationships are preserved in the multilocus case given interlocus additivity, i.e., no epistasis.

The main occupation of the theory is to predict breeding values of individuals, estimate heritabilities of traits, and to predict selection responses.[1,2] Its main methodological tools are matrix algebra, linear models and variance component partitioning. The phenotypic variance (V_P) of a given metric trait is often in textbooks split into additive genetic variance (V_A), dominance genetic variance (V_D), epistatic or interaction variance (V_I) and environmental variance (V_E). Conceptually, modern quantitative genetic theory has not in principle moved beyond what may be developed from a single locus model with two alleles, one dominant and one recessive, in a random mating population.[1,5] The traditional way of developing this mathematical-statistical machinery is to start from the concepts of additive and dominant gene actions, introduce a linear approximation in the form of a least-squares regression of genotypic value on gene content in the single-locus case, define the statistically motivated terms average effect (or breeding value; A) and dominance deviation (D) (Fig. 1a), and from this develop expressions for the variances of A (V_A) and D (V_D), and heritability of the trait, as a function of allele frequency and the degree of dominant gene action (d) (Fig. 1b). This in turn allows establishment of the highly instrumental allele frequency-independent relationship between the performance covariance of relatives and the additive variance V_A (COV(Offspring Parent)=$1/2V_A$) (see ref. 2 for a systematic presentation). By assuming random mating and independent segregation of loci over several generations, the single-locus results are valid for the multilocus case without any further theoretical development. From this foundation, expressions for epistatic variance (V_I) can also be developed.[2,6,7]

In a keynote address for a 1959 Cold Spring Harbor Symposium, Ernst Mayr criticized quantitative genetics models as grossly oversimplified "beanbag genetics", arguing that they naively treated each individual gene as an independent unit (see ref. 8 for a detailed historical exposition). Even though this is not strictly true, it is fair to say that the theory, as it is used within production biology and evolutionary biology, is to a large degree based on the conception that a very large number of alleles are responsible for the observed genetic variation of a trait (the infinitesimal model), and that these alleles behave as discrete elements whose effects can be summed within and between loci. In fact, as long as the effect from each gene is presumed to be small it does not really matter in operational terms whether interactions within and between some of these genes are acknowledged or not. For this reason I think that as long as quantitative genetics theory is based on the infinitesimal model as such, it qualifies to be characterized as beanbag genetics despite that Mayr's reasons for coining the term may be questioned. (O. Kempthorne, one of the founders of quantitative genetics, has in fact also made use of the term to characterize the theory).[9]

Distinctions between Physiological and Statistical Gene Effect Terms

Thirty years ago Lewontin stated with reference to genetic variance components that "an analysis of variance is not an analysis of causes of variability, except in the strictly additive case".[10] This was elegantly exemplified by Cheverud and Routman[11] when they documented the importance of differentiating between physiological and statistical definitions of dominance and epistasis. By making this distinction they were able to provide some refreshing conceptual clarifications and to show that physiological epistasis (the additive or nonadditive gene interaction effects between loci) do indeed contribute to additive genetic, dominance, and epistatic values and variances. This work was apparently partly motivated by the fact that in the classical development the physiological dominance (the nonadditive gene interaction effect between two alleles at the same locus) also contributes to the additive genetic variance.

However, it must be fair to say that the insight of Lewontin is not excessively reflected in the quantitative genetic literature when it comes to interpretation of variance components. According to a key textbook in the field (ref. 1, p. 148, 170), the additive and dominance genetic variances are causal components of variance, despite that it is emphasised in the same book that additive variance can arise from genes with any degree of dominance or epistasis (ref. 1, p. 131). In another key textbook (ref. 2, p. 69) the additive variance V_A is biologically interpreted as the genetic variance associated with the average additive effects of alleles, while the dominance variance V_D is biologically interpreted as the genetic variance associated with dominance effects. Even though this interpretation is statistically valid, it is not very illuminating from a biological point of view, and it has the flavour of an incorrect causal explanation. Furthermore, in quantitative-genetic simulation models the dominance component is frequently associated with dominant gene action only,[12-14] and the additive variance with additive gene action only.[15-17] This implies that there is a tendency to associate V_A causally with additive gene action, and V_D with dominant gene action. However, despite these inclarities the use of quantitative genetic theory has been, and will in many cases continue to be, a highly successful enterprise within animal and plant breeding. This calls for an explanation.

Why Does Quantitative Genetics Work So Well?

I think the main reason for this resides in the fact that when it comes to operational use of quantitative genetic theory within evolutionary genetics and animal and plant breeding, it is the covariance relationships between sibs which are used to estimate heritabilities and selection responses. In his 1989 textbook Falconer (ref. 1, p. 153) stated that the regression of offspring on parents is a useful measure of the degree of resemblance because it is simply related to the causal components of variance. I think this is to turn the validation arrow the wrong direction. The regression of offspring on parents is not only useful; it is the de facto relationship that causes the successful operational use of the heritability concept. By making use of this regression as a basis for defining additive variance and heritability, the theory no longer depends on its own foundation as it is built up from the one-locus case. To see this, assume a linear regression of offspring (O) on midparent value (P). Let O_i and P_i, i=0,1 be the mean offspring and midparent values in generation 0 and 1, respectively. The selection response ($R = O_1 - O_0$) from one generation to the next is then given by

$$R = \frac{\sigma(O,P)}{\sigma_P^2}(P_1 - P_0) = h^2 \Delta S \qquad (1)$$

where h^2 is the heritability of the trait, and ΔS is the selection differential (see Fig. 2 for further explanation). Concerning predictive purposes it is seen from eq. (1) that there is no need for any information about the underlying genetic basis of a metric character (i.e., the number of genes, their degree of polymorphism, and their interaction structure) to make predictive use of the offspring on parent regression.

Thus, the heritability concept is operationally just a renaming of the regression coefficient. This is also the key to understanding the predictive capability of quantitative genetic theory. A positive regression coefficient is a consequence of the fact that offspring resemble their parents. That offspring resemble their parents is a prerequisite for the principle of natural selection to work, which in turn is the main explanatory concept for the phenomenon of adaptation in nature. From this follows that by equating the term heritability with the tendency of offspring to resemble their parents, the operational methodological apparatus of quantitative genetic theory is based upon one of the most generic patterns in biology. Hence, as we do not need to have a proximate understanding of the underlying regulatory mechanisms to make use of this pattern, its predictive capability is not really surprising. However, predictive capability is not synonymous with explanatory capability, and as was clearly seen by Kempthorne,[5] it is not valid to infer that this success is caused by the presence of consistent definitions and concepts with great explanatory and heuristic power at the genetic level. Thus, it would be more precise to call this methodological apparatus quantitative phenetics instead of quantitative genetics, which should be reserved for a theory capable of causally linking genes, phenotypes and the statistical descriptors of population level phenomena. By this I do not mean to say that current quantitative genetics methodology will become totally obsolete within production biology or evolutionary biology. In fact, due to our lack of knowledge it will in some cases be superior to other approaches for a very long time.

The fact that other types of relationship information than the offspring-parent relationship are used in linear mixed animal models to estimate variance components and predict

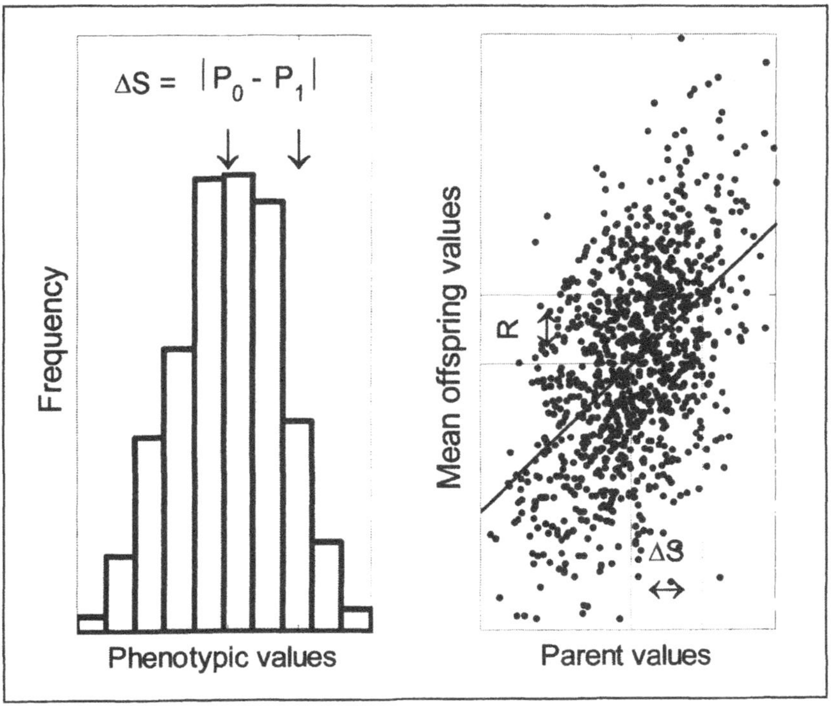

Figure 2. Illustration of how the operational definition of heritability (h^2) links the selection differential (ΔS) and selection response (R) without any need for a biological understanding. The left panel shows the frequency distribution of a parental population of 1000 individuals sampled from a normal distribution ($N(\mu,\sigma)$, μ=10,σ=2). From this population a mean offspring value (y_i) was created for each individual (x_i) according to the equation $y_i=\beta N(x_i,\sigma)$, i=1,...1000 and β=0.3 A new parental population with mean P_1 was made from the 200-highestranking individuals of the parental population, giving the selection differential $\Delta S=P_1-P_0$. The right panel shows the regression of mean offspring value on parent value. From the regression equation $y=\hat{\alpha}+\hat{\beta}x$, the expected mean values (O_0,O_1) of the offspring population without and with selection are then given by $O_0=\hat{\alpha}+\hat{\beta}P_0$ and $O_1=\hat{\alpha}+\hat{\beta}P_1$, respectively. The predicted selection response (R) is then $R=O_1-O_0=\hat{\beta}(P_1-P_0)=h^2\Delta S$.

breeding values (see ref. 2 for an overview and references) does not invalidate the above argumentation. Covariance relationships between siblings, cousins, etc., have their entire basis in the resemblance between offspring and parents. However, in contrast to the simple offspring-parent regression, the coefficients of relationship between relatives in these models reflect that they actually presume a large number of genes to influence a given trait. By this, they do make a presumption about the underlying genetic structure along the lines of classical quantitative genetics theory, but with very modest substantiation in biological facts.

I think it is fair to say that unless we are able to build an explanatory bridge between regulatory biology and the statistical descriptors of population level genetic phenomena, we cannot pretend to have a real theory. It is premature to elaborate on what such a future theory will look like. However, I am quite confident that it will have the following four characteristics: (i) its explanatory apparatus will become based on the regulatory principles observed in biological systems; (ii) it will be capable of explaining population level genetic phenomena with reference to these regulatory principles; (iii) the mathematical language of nonlinear system dynamics will become a core element of the mathematical foundation of this theory; (iv) in stark contrast to current practice within (natural) science, the explanatory apparatuses of nonlinear system dynamics and mathematical statistics will become highly integrated.

The development of such a theory will keep theoreticians and experimentalists busy for a very long time. As in every other aspect of life where we want to master something difficult, we have to start playing with something simple. Compared to most other population level genetic phenomena, genetic dominance is a simple one. We need in principle only three individuals to document it, it is in principle a single locus feature, it is an emergent nonlinear property of gene regulatory systems, and we have a reasonable understanding of at least some of the molecular mechanisms likely to be involved. In the following I will show you that it is possible to establish a causal basis of genetic dominance by translating relevant molecular biological information into the vocabulary of nonlinear system dynamics.

The Implicit Connection between Genetic Dominance and the Feedback Concept

Quantitative genetics theory considers genetic dominance to be an intralocus interaction phenomenon.[1,2,4] Interpreted in modern molecular biological terms the main actors generating genetic dominance are thus the two alleles, their mRNA products and their protein products embedded in a supportive cellular environment. Considering what is the major mechanism of transcriptional regulation (see e.g., the textbook by Lewin),[18] it follows that in most intralocus genetic dominance situations the protein products from the alleles will have to influence the expression pattern of the locus. Thus, if we take our textbook molecular biology seriously, the classical definition of genetic dominance implies the existence of an autoregulatory feedback loop. In fact, ubiquitous existence of such loops is now well documented. In *Escherichia coli* genetic regulatory networks it seems that more than 50% of the genes encoding transcriptional regulators are autoregulated,[19,20] while in yeast the number is estimated to be about 10%.[21] We do not have such figures for higher eukaryotes yet, but there are numerous reports of autoregulatory loops also in this case.[22-26] We also have to remember that very many genes seem to be regulated by interactions with multiple transcription factors binding to their *cis*-regulatory systems. When a gene is to be expressed or repressed in a number of different contexts, it is frequently found that separate *cis*-regulatory elements carry out different parts of the overall regulation.[27] This suggests that many genes in higher eukaryotes may show autoregulation as well as other types of regulation depending on the spatio-temporal regulatory situation.

Because widely differing definitions of feedback abound in the literature, and because many use the term uncritically without having a strict definition in mind at all, it is timely to say a few words about the concept before we proceed with the models.

The basic ingredient in feedback is the effect, widely speaking, of one state variable on another. I prefer to use the term "action" for such effects, to emphasize that it is a directed relationship, which may or may not be reciprocated.[28] A definition of "action" widely used in theoretical regulatory biology is that x acts positively on y if an increase in x would modify the rate-of-change of y in the positive direction, and vice versa.[29] A feedback loop is then a closed chain of such actions among state variables. It is called a positive feedback loop if there is an even number of negative actions in the chain, and a negative feedback loop otherwise. Given that the actual system is described by a set of ordinary differential equations, the action of x on y is positive if the partial derivative $(\partial/\partial x)(dy/dt)$ is positive, and vice versa. All the feedback loops in operation at a specific state can be read out from the Jacobian matrix of the system (see Box 1 for further explanation).

The above definition of "action" is the basis for several results on the correspondence between the feedback structure of a system and its dynamical properties. For instance, positive

Box 1. Illustration of how to calculate the feedback structure of an ordinary differential equation system at a given point in state space.

Given a system of ordinary differential equations:

$$\frac{dx_1}{dt} = F_1(x_1, x_2, x_3),$$
$$\frac{dx_2}{dt} = F_2(x_1, x_2, x_3),$$
$$\frac{dx_3}{dt} = F_3(x_1, x_2, x_3),$$

Develop the general Jacobian (*J*) of the system:

$$J = \begin{pmatrix} \frac{\partial F_1(x_1,x_2,x_3)}{\partial x_1} & \frac{\partial F_1(x_1,x_2,x_3)}{\partial x_2} & \frac{\partial F_1(x_1,x_2,x_3)}{\partial x_3} \\ \frac{\partial F_2(x_1,x_2,x_3)}{\partial x_1} & \frac{\partial F_2(x_1,x_2,x_3)}{\partial x_2} & \frac{\partial F_2(x_1,x_2,x_3)}{\partial x_3} \\ \frac{\partial F_3(x_1,x_2,x_3)}{\partial x_1} & \frac{\partial F_3(x_1,x_2,x_3)}{\partial x_2} & \frac{\partial F_3(x_1,x_2,x_3)}{\partial x_3} \end{pmatrix}$$

The Jacobian matrix is a mathematical tool common to most published treatments of feedback. Its elements J_{ij} are the partial derivatives of the rate-of-change of variable x_i with respect to variable x_j, and they express the action of x_j on F_i. Note that the signs and values of the various elements (or actions), and thereby the feedback loop structure, may vary all over state space.

Calculate the partial derivatives and find the action signs at a specific point $\overline{X}_0$:

$$A = \begin{pmatrix} - & 0 & + \\ - & + & + \\ - & + & - \end{pmatrix}$$

The point $\overline{X}_0 = (x_1, x_2, x_3)$ may for example be a stationary point where $dx_i/dt = 0, i = 1,2,3$, but it can also be any other point in the state space of the system.

Read out the feedback loops from the action sign matrix (*A*):

Negative autoregulation: $A_{11}(1 \xrightarrow{-} 1)$ and $A_{33}(3 \xrightarrow{-} 3)$
Positive autoregulation: $A_{22}(2 \xrightarrow{+} 2)$
Negative 2-element loop: $A_{13}(3 \xrightarrow{+} 1)$, $A_{31}(1 \xrightarrow{-} 3)$
Positive 2-element loop: $A_{23}(3 \xrightarrow{+} 2)$, $A_{32}(2 \xrightarrow{+} 3)$
Negative 3-element loop: $A_{13}(3 \xrightarrow{+} 1)$, $A_{21}(1 \xrightarrow{-} 2)$, $A_{32}(2 \xrightarrow{+} 3)$

feedback is required for the existence of multiple stable points,[30-32] and a negative feedback loop of length two or more is required for a stable periodic solution.[31,32] (See Vik and Omholt (subm.) for a more detailed analysis of the feedback concept).

One may say that regulatory biology is to a considerable extent a story about negative and positive feedback processes. Negative feedback is normally associated with maintenance of a given state (homeostasis), while positive feedback is normally associated with differentiation from one state to another. This is likely to be too simplistic, and probably we still have some way to go before we have a good overview of which types of regulatory situations the two feedback processes are separately involved in, and where they both participate. But we know at least that life is completely dependent upon their interplay.

In the following I will show that the feedback concept is not only causally linked to the phenomenon of genetic dominance, but that some generic features of genetic systems can actually be predicted from their feedback structures.

Describing and Analysing the Dynamics of an Autoregulatory Locus

Description of the Model

Let a gene be defined as a structural unit composed of a regulatory region and a functional region. The regulatory region includes all the DNA of the gene that either directly or indirectly is important for transcriptional, posttranscriptional, translational, and posttranslational control of the functional protein. The functional region of the gene includes all DNA that influences the actual function of the protein product. Furthermore, assume that the protein made from the nonpolymorphic functional region of each allele binds monomerically or multimerically to the two polymorphic regulatory regions. Finally, assume that in the positive autoregulatory case there is no polymorphism in the additional regulatory region and in the regulatory gene(s) controlling the initial onset of the production of the protein product (which is operative until the gene product has reached a concentration by which it can stimulate its own production).

The situation with intralocus regulatory interaction can then be described by a differential equation system expressing the time rate of change of the protein product concentrations x_1 and x_2 from two alleles located at the same locus X,

$$\dot{x}_1=\alpha_1 R_1(y)-\gamma_1 x_1,$$
$$\dot{x}_2=\alpha_2 R_2(y)-\gamma_2 x_2, \qquad (2)$$

where $y= x_1+x_2$ is the total gene product concentration, α_1 and α_2 are the maximum production rates, $\gamma_1>0$ and $\gamma_2>0$ are the relative degradation rates, and R_1 and R_2 $(0\leq R_j\leq 1)$ are the production regulatory functions for the two alleles of the gene. R_1 and R_2 are assumed to be continuous and differentiable functions of y and all their parameters. This model is a "diploid" version of the "haploid" gene regulatory models previously investigated in detail by several workers.[33-37] If the two equations have identical parameters and rate functions, the system describes a functional homozygous locus. If at least one parameter or the function in the first equation is different from the corresponding parameter or function in the second equation, the system describes a heterozygous locus. If R_j, j = 1, 2 are monotonically decreasing, eqs. (2) represent a one-locus model with negative feedback. If R_j, j = 1, 2 are monotonically increasing, eqs. (2) represent a one-locus model with positive feedback.

The equilibrium value of the protein product from each allele (x_1^* and x_2^*) is found by solving the equations $\dot{x}_1=\dot{x}_2=0$(eqs. (2)) with respect to x_1 and x_2. From this we get $x_1^*=(\alpha_1/\gamma_1)R_1(y^*)$ and

$x_2^*=(\alpha_2/\gamma_2)R_2(y^*)$. Remembering that $y^*=x_1^*+x_2^*$, we see that the locus can display the three different equilibrium values or phenotypes (y^{11},y^{12},y^{22}) given by

$$\begin{aligned} y^{11} &= x_1^* + x_1^* = 2\frac{\alpha_1}{\gamma_1}R_1\left(y^{11}\right), \\ y^{12} &= x_1^* + x_2^* = \frac{\alpha_1}{\gamma_1}R_1\left(y^{12}\right)+\frac{\alpha_2}{\gamma_2}R_2\left(y^{12}\right), \\ y^{22} &= x_2^* + x_2^* = 2\frac{\alpha_2}{\gamma_2}R_2\left(y^{22}\right). \end{aligned} \tag{3}$$

The equilibrium values y^{11} and y^{22} are the phenotypes of the two homozygous genotypes, and y^{12} is the phenotype of the heterozygous genotype. From the phenotypic values y^{11},y^{12},y^{22} the degree of physiological dominance (*d*) for this locus may be found by traditional means. Following Falconer it is given by,

$$d=\frac{y^{12}-y_m}{\left|y^{22}-y_m\right|}, \text{ where } y_m=\frac{y^{11}+y^{22}}{2}$$

is the midpoint between the homozygotes.[1] When, $d=0$, the locus is said to show additive gene action (additivity), when $0<|d|<1$ it shows negative or positive partial dominance, when $|d|=1$ it shows negative or positive complete dominance, and when $|d|>1$ it shows negative or positive overdominance.

Omholt et al[38] investigated these intralocus models in the range between two extreme regulatory situations characterized by all regulatory interactions being based on a switch-like effect-response mechanism and an ordinary Michaelis-Menten mechanism, respectively. The very useful Hill function

$$S(y,\theta,p)=\frac{y^p}{\theta^p+y^p}$$

was used as the regulatory function.[39] When, $p=1$, it describes an ordinary hyperbolic Michaelis-Menten function, when $p=2$ is starts to show sigmoidal form, and when $p\to\infty$, S approaches the unit step function with threshold θ. Negative autoregulatory functions can then be exemplified by

$$R_j(y)=1-S(y,\theta_j,p_j),\ j=1,2, \tag{4}$$

while positive autoregulatory ones can be exemplified by

$$R_j(y)=S(y,\theta_j,p_j),\ j=1,2. \tag{5}$$

Figure 3 depicts the two types of regulatory functions together with their associated network structures in the autoregulatory case.

Dominance Patterns of an Autoregulatory Locus

In the negative autoregulatory case Omholt et al[38] showed that eqs. (2) have a single stable state for each of the three genotypes. Dominance is the rule, and the degree of dominance *d* varies as a function of parameter values. Overdominance never occurs. Additive gene action is only observed when the negative feedback loop is not activated for any of the genotypes, and there is only constitutive expression and no intralocus interaction. Additivity is thus not a characteristic of the negative autoregulatory loop.

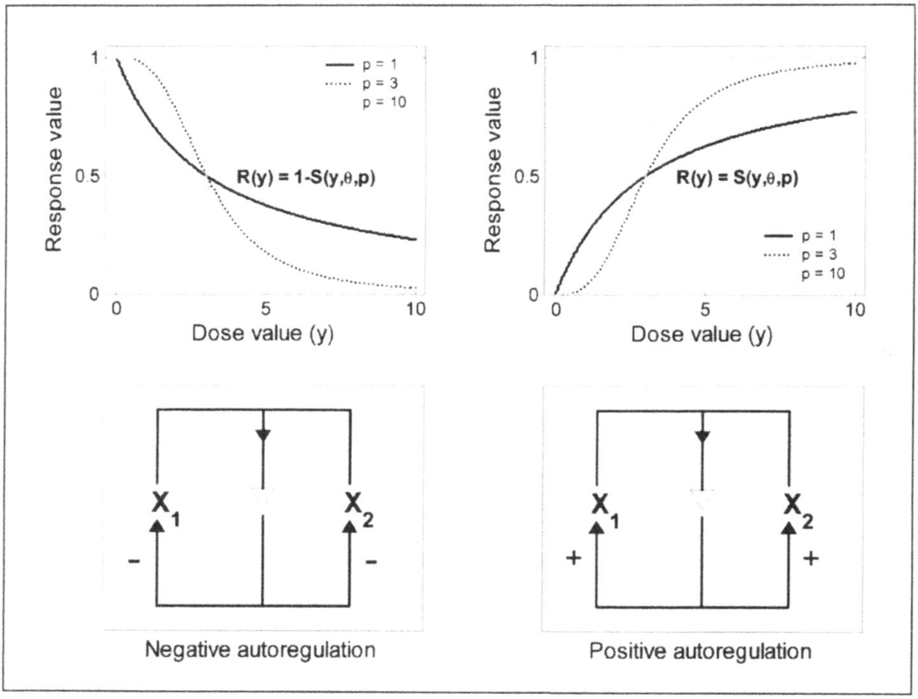

Figure 3. The two types of regulatory functions and their associated action diagrams in the intralocus interaction case. The two columns show various forms of the regulatory function and the associated action diagram in the negative and positive autoregulatory case, respectively. X_1 and X_2 represent the two alleles of a gene X. The triangle indicates that the concentrations of the gene products x_1 and x_2, from allele 1 and allele 2 at locus X, are added and that the total sum $y=x_1+x_2$ regulates the activity of the gene by binding to the regulatory region of X. The two upper panels show how suitable the Hill function $S(y,\theta,p)=y^p/(\theta^p+y^p)$ is for modelling various regulatory situations. By just changing the parameter p the form and the steepness of a regulatory relationship can be changed dramatically (θ=3 in both cases, and note that $S(\theta,\theta,p)$=0.5). For clarity, the negative loops representing the decay terms of the protein products from each allele have been left out in the lower panels. Note that these decay terms contribute to feedback in the direction of negative feedback.

In contrast, in the positive autoregulatory case additive gene action is the prevalent pattern. However, there is also ample room for dominant as well as overdominant gene action. Furthermore, for each of the three genotypes there will in general be several equilibrium states, some of which are stable. This may be an important feature, as it implies that a locus may display completely different gene action patterns depending on the perturbation history of its associated mRNA or protein product concentrations, or just internal noise in gene expression.[40] The phenomenon illustrates a deep general fact about dynamic systems conjectured by Rene Thomas and first proved by Erik Plahte,[30] namely that in dynamic systems having two stable states there must be at least one positive feedback loop. The heuristic potential of this theorem within regulatory biology is considerable because it helps us to unravel the interaction structure between underlying systemic units from the systems behaviour.

If a protein or an mRNA expression pattern of a diallelic locus shows a nonadditive arrangement, it would be tempting to attribute genetic dominance behaviour to the locus. However, in the following section we will see that one must be very careful when making interpretations from such data. Dominance patterns may actually be caused by epistasis, which in turn

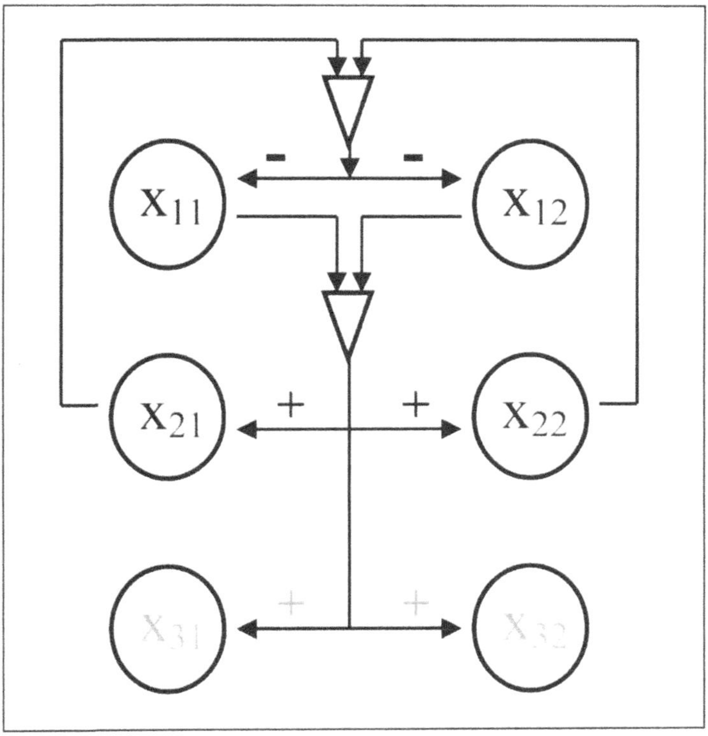

Figure 4. Interlocus model. Two possibly polymorphic loci X_1 and X_2 interact by their gene products through a negative feedback loop, and X_1 positively regulates the possibly polymorphic locus X_3. The triangles indicate that the outputs of the gene products x_{i1} and x_{i2}, $i=1,2$, are added and that the total sums $y_i=x_{i1}+x_{i2}$ regulate the activities of the three genes. For clarity, the negative loops representing the decay terms of the protein products from each allele have been left out.

suggests that dominance and epistasis phenomena are more closely linked than what has previously been acknowledged.

Relationships between Genetic Dominance and Epistasis

Description of the Model

To include the phenomena of epistasis and downstream regulatory effects in a feedback structure we need a three-locus model, like for example the one depicted in Figure 4. One mathematical representation of this structure is:

$$\dot{x}_{1j}=\alpha_{1j}[1-S(y_2,\theta_{1j},p_1)]-\gamma_{1j}\,x_{1j},$$

$$\dot{x}_{2j}=\alpha_{2j}S(y_1,\theta_{2j},p_2)-\gamma_{2j}\,x_{2j}, \qquad (6)$$

$$\dot{x}_{3j}=\alpha_{3j}S(y_1,\theta_{3j},p_3)-\gamma_{3j}\,x_{3j},$$

where $y_1=x_{11}+x_{12}$, $y_2=x_{21}+x_{22}$, and $j=1,2$.[38] Equations (6) describe the time rate of change of the concentrations of the protein products, from the three genes X_1, X_2 and X_3 and their alleles 1 and 2, respectively. It exemplifies a very simple three-locus signal transduction network with two loci X_1 and X_2 connected in a negative feedback loop (illustrating mutual causality), and one down-stream locus X_3 activated by the summed protein product concentration y_1 of allele 1 and 2 of locus X_1 only. Except for some additional indexing to keep track of the various loci and alleles, the mathematical structure of eqs. (6) is similar to the structure of the previous single locus model (eqs. (2)). The equilibrium values of the protein products are easily calculated by use of a standard numerical simulation package for a given set of parameter values.

Dominance Patterns of the Three-Locus Model

It is important to note that there are no intralocus interactions in eqs. (6). Thus, according to the classical definition of dominance this model cannot generate any biologically based genetic dominance pattern. This implies that the expression patterns predicted by the model are due to interlocus interactions and actions, i.e., epistasis.

Let us first consider the two loci X_1 and X_2. A mathematical analysis shows that there are only unique and asymptotically stable solutions for y_1 and y_2 for all three genotypes,[38] which means that we in this case do not have such things as for example periodically varying protein phenotypes or protein phenotypes that spend a long time to get back to their equilibrium values after perturbations. When only one of the loci is polymorphic, this induces additive as well as negative and positive dominant gene action expression patterns for both loci. With both loci polymorphic all gene action patterns can be realized, including overdominance, for both loci. However, the overdominance will be present in only one locus at a time. The less steep the functional regulatory relationships become, the less complete will the dominance be.

The fact that dominance appears to be present at the X_2 locus even if only the X_1 locus is polymorphic is probably of some biological relevance. In biological terms the presence of such an apparent dominance pattern means that if the X_1 locus is polymorphic and the y_2 protein product concentration is assayed, one would observe that the X_2 locus showed dominance. Attributing this to intralocus interaction in X_2 would be erroneous. If the y_2 product concentration segregated as a simple Mendelian character, any search for causative genetic variation among polymorphisms in the regulatory or the coding region of X_2 would be futile. Such dominance generated by feedback and polymorphism in another gene in the loop, may be called *epistatic feedback-mediated genetic dominance* (Fig. 5).[38]

Now consider the expression pattern of the protein product y_3=x31+x32 of the downstream locus X_3. When X_1 and X_2 are nonpolymorphic, polymorphism in X_3 results in strictly additive behaviour independent of the degree of steepness of the regulatory effect-response relationships involved. With upstream genetic variation, the downstream locus may show dominance as well as overdominance due to epistasis even if it is nonpolymorphic. This implies that a dominance effect may be mediated systematically through a number of other loci in a regulatory network. This type of epistasis, which in principle does not need to involve any feedback structures, may be called downstream-mediated epistatic genetic dominance (Fig. 5).[38] In biological terms this indicates that a polymorphic regulatory locus high up in a hierarchy (e.g., a transcription factor) may generate heritable dominance effects through epistasis in numerous loci coding for structural gene products realizing metric characters. Search for causative variation among the polymorphisms of these downstream loci would then be futile also in this case. In fact, the report by Damerval and De Vienne[41] on how a transcription factor mutation in maize (*opaque-2*) affected the polypeptide expression pattern seems to provide some encouraging support for the presence of *downstream-mediated epistatic genetic dominance.*

Considering the ubiquity of hierarchy and feedback in regulatory networks[42] it is likely that these two types of epistasis phenomena will be frequently encountered, and that the above distinctions may be helpful when interpreting mRNA and protein expression levels of specific candidate genes within biomedicine as well as animal and plant breeding.

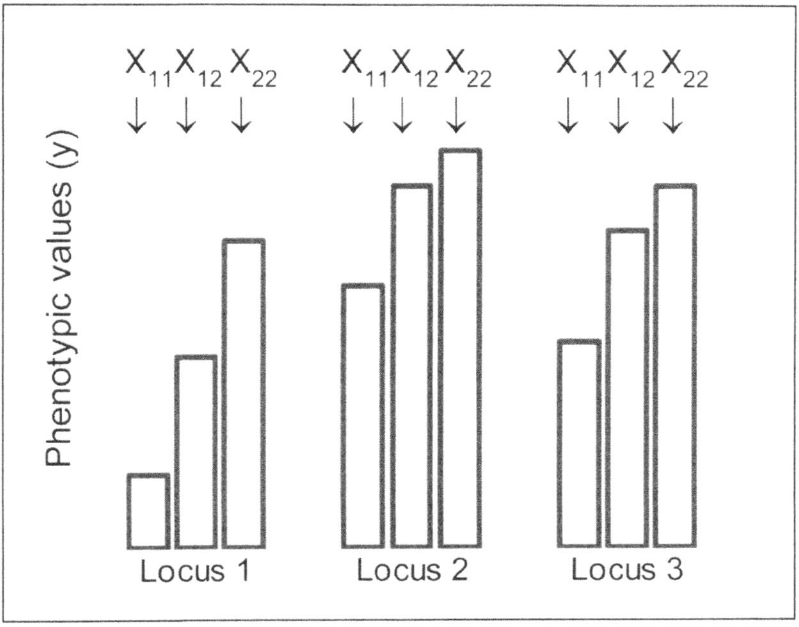

Figure 5. Predicted relationships between epistasis and dominance. The underlying dynamic system is given by eqs. (6). The equilibrium expression patterns of the three loci are based on polymorphism at the X_1 locus only, i.e, the two other loci are homozygous. The connections between the three X_1 genotypes and the phenotypic expressions patterns of the X_2 and X_3 loci show that an apparent dominance effect can segregate as a simple Mendelian character at nonpolymorphous loci due to epistasis. Note that in this particular case the polymorphism at X_1 causes additive behaviour at the X_1 locus itself, and apparent dominance behaviour at the two other loci. Parameter values underlying the numerical simulation: $\alpha_{11}=100$, $\alpha_{12}=200$, $\alpha_{21}=\alpha_{22}=100$, $\theta_{11}=20$, $\gamma_1=\gamma_2=\gamma_3=1$, $\alpha_{31}=\alpha_{32}=100$, $\theta_{12}=80, \theta_{21}=\theta_{22}=25$, $\theta_{31}=\theta_{32}=40, p=1.0$.

Then what about additive gene action in such signal transduction pathways? In the given example, when the lowest equilibrium value of y_1 is much greater than the threshold concentration where y_1 turns on the production of y_3, all functional allelic variation at the X_3 locus influencing the production rate of y_3 will result in approximately additive behaviour. This follows directly from the equilibrium expressions of the two last equations in eqs. (6) if we let $S(y_1, \theta_{3j}, p) \approx 1, j=1,2$. This feature of the downstream locus shows how additive gene action can in fact be a generic property of highly nonlinear hierarchic regulatory networks (see ref. 41 also on this point).

Further Qualification of the Dominance Concept

One might object that the concepts dominance and epistasis are not used properly above by introducing the terms epistatic feedback-mediated genetic dominance effect and epistatic downstream-mediated genetic dominance effect, and that we should restrict the use of the physiological dominance concept to the case of intralocus autoregulatory interaction only. However, the results by Omholt et al[38] imply that there are likely to be many phenotypic patterns attributed to genetic dominance that are not based on intralocus interactions. If one restricts the use of the dominance concept to intralocus interaction one would have to rename dominance patterns as nonadditive patterns. There would then be nonadditive expression patterns caused by intralocus positive or negative interactions that might be called genetic dominance patterns, and nonadditive expression patterns caused by interlocus negative or positive feedback interactions or interlocus downstream positive or negative actions. In this way one could not use the dominance concept properly without access to a molecular genetic knowledge that is at present beyond reach in most cases. By introducing terms linking dominance and epistasis one can use the dominance concept in a consistent way, while at the same time recognizing that genetic dominance may have an intralocus as well as an interlocus basis. In addition, these concepts also contribute to the clarification of the relationship between dominance and epistasis as well as a much-needed qualification of the epistasis concept.[43,44]

An extensive study by Leonardi et al[45] illustrates the above considerations very well. Using two-dimensional polyacrylamide gel electrophoresis (2-D PAGE), the genetic variation of proteins was investigated in three organs (mesocotyl, sheath and blade of second leaf) from two maize lines and their hybrids. Out of the 986 spots numbered over the three organs, 155 polypeptides displayed interline variations in at least one organ. Of these 155 variable spots, 136 showed a hybrid spot intensity halfway between the two parental spot intensities (i.e., additive inheritance), and 19 showed a hybrid spot similar to one of the parental spots (i.e., complete dominance). The authors characterized the nonadditive patterns as dominance patterns, while they explicitly acknowledged that the observed variation could be due to allelic polymorphism of structural genes and/or to polymorphism of any genetic elements controlling protein amounts. I think this reflects a sound practice because we can qualify the observed patterns when we obtain more detailed information about the regulatory structures involved.

Another most noteworthy observation in the report by Leonardo et al[45] was that in all cases but one the dominant inheritance was organ-specific. This certainly tells us something about the need for qualifying the link between epistasis and dominance, but it also tells us something about the need for qualifying the placticity of gene networks concerning their capacity for generating additive and dominant gene action patterns. Concerning the last point it is encouraging to observe that, with monotonic regulatory functions, the simple model above predicts that any degree of dominance displayed by the X_1 locus can be accentuated or even reversed, depending on choice of parameters in the three regulatory functions in eqs. (6).[38] In biological terms this means for example that a single transcription factor can orchestrate a bewildering variety of epistatic downstream-mediated genetic effects depending on the structure of the regulatory elements of the downstream genes, and that spatiotemporal regulation of the transcription factor itself by other genes may lead to organ-specific variation and inheritance.

Where Do We Go from Here? Outline of a Possible Research Program

By acknowledging that genes interact in feedback structures, and making use of an appropriate mathematical language to describe and analyse such genetic structures, a whole new research field within the disciplinary frames of quantitative genetics opens up. In the following I outline some of the tracks that can be followed.

Sophistication of Diploid Regulatory Models

The above models are relevant for intra- as well as intercellular regulatory systems, and they are sufficiently complex to catch most of the molecular mechanisms suggested by Wilkie[46] to be responsible for generating dominant mutations, such as reduced gene dosage, expression or protein activity (haploinsufficiency), increased gene dosage, ectopic or temporary altered mRNA expression, increased or constitutive protein activity, and dominant negative effects. However, there is much need for a systematic search for generic patterns in regulatory structures including metabolic nets concerning their capacity for generating additivity, dominance and epistasis.

The data[21,42] we are starting to get about which regulatory structures are actually realised in eukaryotic systems at the transcriptional level, suggest that a considerable amount of this work, when it comes to making quantitative predictions, can be done by use of nonlinear differential equations. However, it is likely that the differential equation approach is not suited for describing all mechanisms creating dominance and epistasis phenomena. In some of these cases, we might even need mathematical methodologies not yet identified or even developed to describe certain types of regulatory networks.

Relationships between Variance Components and Regulatory Structures

Even though I have tried to give an explanation of why quantitative genetics works so well with reference to its exploitation of the covariance structure between relatives, it remains to be explained how and when gene regulatory networks and signal transduction pathways, with all their nonlinear interactions and hierarchical organization, promote statistical additivity and nonadditivity.

An almost totally unexplored field is how various regulatory structures contribute to the various statistical terms estimated and used in quantitative genetics (V_A, V_D, V_I) as a function of allele frequencies. We know for example that physiological dominance and epistasis contribute to V_A, but we are far from having an understanding of this in terms of first principles of how genes work and interact. To find generic relationships between genetic regulatory architectures and statistical descriptors of populations is an important step in the establishment of a real quantitative genetic theory connecting the behaviours of genes to the statistical descriptors of genetic variation at the population level. Though this task might seem daunting, quantitative trait loci (QTL) data[47-52] showing that rather few factors appear to be responsible for a considerable portion of observed phenotypic variance in animals and plants are encouraging. They suggest both that such a theoretical development is within reach, and that the infinitesimal model is in many cases blatantly wrong. Current quantitative genetics does not distinguish between the number of genes responsible for the expression of a metric trait, and the number of genes actually carrying functional polymorphisms influencing the observed genetic variation of the trait. In contrast, variation at a single gene may according to the feedback genetics paradigm generate a lot of variation, while the rest of the genes underlying the trait are not polymorphic (see Thoday and Thompson[53] for a very early and clearheaded presentation of the small number of genes that are actually necessary for producing phenotypic distributions that are indistinguishable from normal distributions frequently thought to be caused by many loci).

Another important aspect in this connection is that as soon as we open the black box of the individual, the phenotypic variance attributable to the environment can be studied and understood much more properly in the context of dynamic models than statistical ones. In fact, the genotype x environment interactions frequently found by statistical analyses are something that should be explained in mechanistic terms by a future real quantitative genetics theory.

Maintenance of Genetic Variation

The classical genetic approach predicts that strong directional, and to some degree stabilizing, selection usually erodes additive genetic variance.[54-56] Gillespie and Turelli stated in 1989 that "the mechanisms responsible for genetic variation in quantitative traits in natural populations remain a mystery".[57] Despite numerous contributions dealing with this subject, maintenance of genetic variation under (strong) selection regimes still seems to be an enigma for quantitative genetics theory. As soon as we include feedback processes in our thinking, maintenance of genetic variation is no longer such a mystery.

We have seen that dominance is a generic feature of negative autoregulatory loci, and that dominance effects can be propagated and show up at downstream nonpolymorphic loci in

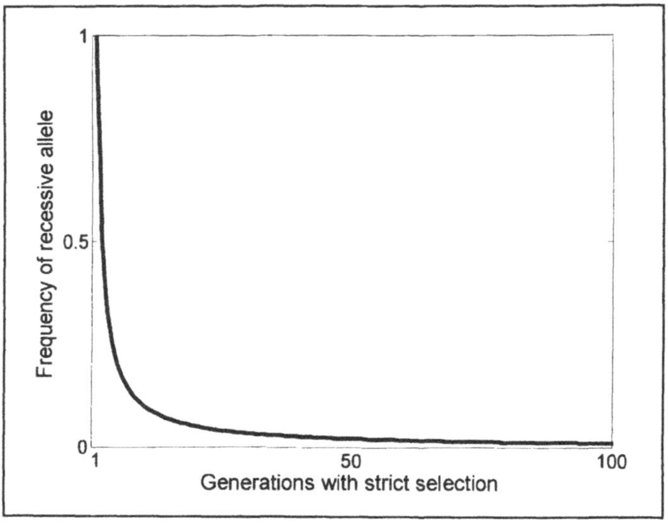

Figure 6. The fate of a recessive allele when all recessive homozygotes are removed in each generation. Presuming Hardy-Weinberg conditions, the allele frequency of the recessive allele in generation i (q_i) was calculated from the recursive formula $q_i = 1-(p_{i-1}^2 + p_{i-1}\, q_{i-1})/(1-q_{i-1}^2)$ which after some algebraic manipulation gives $q_i = q_0/1+iq_0$, where i=1,...100, $q_0=0.99$, and $p_0=1-q_0$.

regulatory nets. Thus it is worth first by a classic genetic approach to consider what happens when such a locus effecting dominance is exposed to strong selection. Let us assume that the locus has only two alleles, and that the heterozygote resembles one of the homozygotes so much that natural selection is not capable of distinguishing the two genotypes from their phenotypic signatures. Further assume that all the recessive homozygotes are removed in each generation. Under Hardy-Weinberg conditions, the recessive allele will still be present in the population after 100 generations (Fig. 6). The biological lesson from this simple exercise is that recessive alleles at loci generating dominance effects are very likely to be present in the population unless genetic drift wipes them out by chance. This latent variation may then be exploited later on by change in natural or artificial selection regimes.

Regulatory nets are capable of generating latent variation in much more subtle ways than the previous example.[58-60] It is beyond the scope of this chapter to make a detailed treatment of this, but I would like to make a simple illustration of a deep generic principle of dynamic systems related to maintenance of genetic variation discovered by Plahte et al.[37] Assume that two genes are connected by a negative feedback loop like X_1 and X_2 in eqs. (6). By running this system on the computer for an initial parameter set we will get a distinct phenotype defined by the equilibrium values of y_1 and y_2. Now let us assume we run this system 50 more times with production rate values sampled randomly from the intervals $\alpha_{1j} \pm 0.5\alpha_{1j}$ and $\alpha_{2j} \pm 0.5\alpha_{2j}$, j=1,2, and keeping all other parameters fixed. The location of these 50 new equilibrium values in the y_1, y_2-space reflects the phenotypic variation associated with these 50 genotypes. If we do this exercise twice where we either let all regulatory interactions become quite steep sigmoidal (p = 10) or hyperbolic (p = 1), we find that the phenotypic variation differs dramatically in the two cases (Fig. 7). This is an important result, because it suggests that genetic nets with switch-like dose-response regulatory relationships may possess huge amount of potential functional genetic variation that is not visible to natural selection before the appearance of a key mutational event changes the interaction structure as such. The accumulated variation may then cause

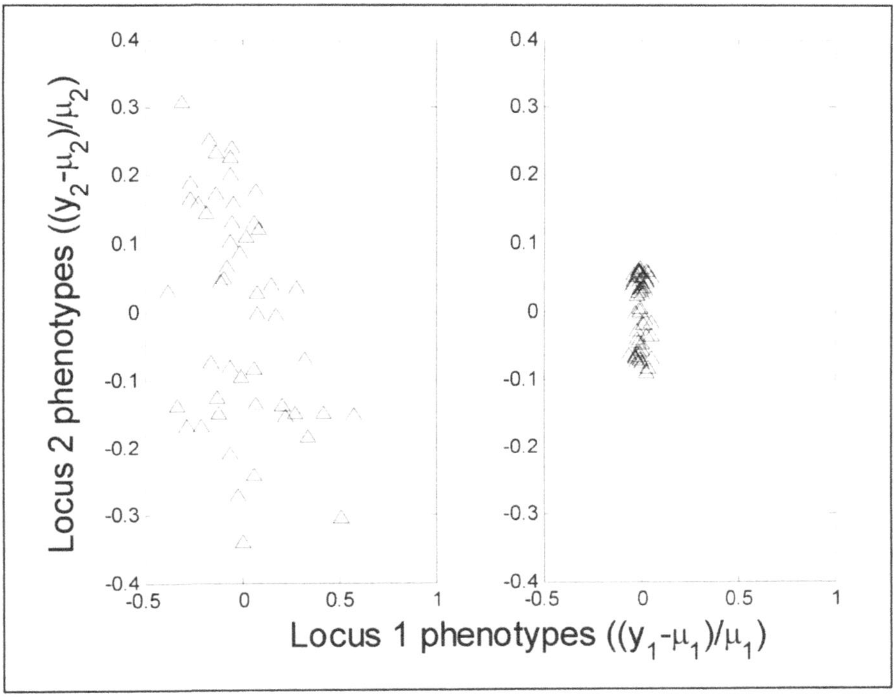

Figure 7. Illustration of systemic silencing of functional genetic variation. Two loci are connected by a negative feedback loop like X_1 and X_2 in eqs. (6). The plots show the distributions of the equilibrium values of the protein products from the loci for 50 separate runs where the production rate values were sampled randomly from the intervals $\alpha_{1j}\pm0.5\alpha_{1j}$ and $\alpha_{2j}\pm0.5\alpha_{2j}$, j=1,2. The two numerical simulations underlying the left and right panels differ only by the steepness of the regulatory functions. Left: p=1 (hyperbolic, Michaelis-Menten type), right: p=0.1 (steep sigmoidal). All other initial parameters were equal: $\alpha_{11}=\alpha_{12}=\alpha_{21}=\alpha_{22}=100$, $\gamma_1=\gamma_2=1$, $\theta_{21}=40$, $\theta_{22}=20$, $\theta_{11}=\theta_{12}=25$.

distinct phenotypic changes and thus contribute to V_G. I call mutations of this type systemically silenced mutations. Steep sigmoidal dose-response relationships are thought to be a common characteristic of transcriptional systems as a result of the cooperative recognition of promoters/enhancers by transcription factors and/or their synergy to attract the basal transcriptional machinery.[61] I am therefore quite confident that we will find this kind of systemic silencing in many situations (see Xiang et al (subm.) for a detailed theoretical treatment of this type of robustness in regulatory nets).

If we take into account the incredibly complex regulatory nets that prevail in organisms, it follows from the above considerations that maintenance of genetic variation as well as long-term selection response may actually be generic features of regulatory networks. However, there is need for numerous theoretical and experimental studies before we get a firm grasp of the processes underlying these phenomena.

Genetic Dominance and Life History Strategies

Since the contributions by Fisher[62-66] and Wright[67-69] there has been a debate whether the phenomenon of genetic dominance needs an evolutionary explanation or not.[70-75] Omholt et al[38] provided reasons for why the opposing views, but not necessarily the actual explanations given, might be reconciled to some degree by use of the feedback genetics paradigm. The debate is by no means over, and as we are finally getting access to proper experimental tools to probe this issue, I find it timely to give a brief outline of some personal viewpoints here.

Genetic dominance phenomena are generic features of regulatory and metabolic nets. However, the designs of nets displaying dominance may have been selected for other reasons than their capacity for generating dominance phenomena. What we really would like to know, and which already so much mental energy has been spent on, is how genetic dominance as such is treated by natural selection. Genetic dominance patterns may be selected against, they may be selectively neutral, or they may be selected for. If all dominance patterns were selected against, there would be no direct evolutionary reason for their presence. However, if they are selectively neutral or are selected for, we are in need of specific explanations.

First of all, a considerable amount of nonadditive gene actions in the physiological sense underlying a metric trait is likely to end up in the additive variance of the trait.[11] Directional selection can thus exploit this variation directly, and the dominance patterns are selected for along with additive gene action patterns. However, even though this is an important issue (see section *Relationships between Variance Components and Regulatory Structures*), this is not my main concern here. Rather it is that nonadditive gene actions are likely to generate more variation among the progeny than additive gene actions at certain allele frequencies (compare *VG* and *Vg* in Fig. 1b for a very simple illustration), and that increased variation among the progeny may under certain life history conditions be selected for.

Crnokrak and Roff[76] found that traits for wild species closely associated with fitness (life history) had significantly higher dominance components than did traits more distantly related to fitness (e.g., morphology). This is in accordance with quantitative genetics theory predicting that selection usually erodes only additive genetic variance while not affecting dominance variance.[54-56] However, Merila and Sheldon[77] showed that fitness-related traits tend to have low heritabilities compared to nonfitness traits not because of lower additive variances (indeed they tend to be higher) but rather because of much higher nonadditive variances. Thus the available empirical data suggest that the high genetic dominance variances of the fitness characters in wild species may be caused by selection for dominant gene actions per se and not something that is left after most of the additive variance due to additive gene action has been eroded.

How can the observations of Merila and Sheldon be explained in evolutionary terms? There may in fact be several evolutionary explanations that are not mutually exclusive. However, for obvious reasons I like to call attention to the following one by Omholt et al.[38] It provides a testable life-history explanation of why and when natural selection would actually favour dominant gene action patterns compared to additive gene action patterns. The argument goes as follows: Individuals in sexually reproducing species characterized by production of a small number of young progeny per lifetime want to centre their progeny more narrowly about the parent value than individuals producing a large number of young progeny during their lifetime. On average, irrespective of the life history situation, only a couple or so progeny per lifetime per female enter the reproductive cycle. If the organism produces large numbers of progeny that later on are exposed to an unpredictable abiotic and biotic environment, it may be more advantageous to let the progeny spread considerably about the parent value instead of keeping them narrowly centred. In contrast, a low-progeny producer may be better off following the simple mantra: "I have been able to make it so far, which means that I am probably quite fit, and therefore should stay away from gambling having offspring too much different from myself". In less antropomorphic terms this can be achieved by selection against nonadditive gene actions in the genetic structure underlying the given trait through lowered fitness of parents having more variable progeny.

There may be several situations where a regulatory structure might realize a dominance pattern or an additive pattern and that there exist broad parameter domains (i.e., allelic variation domains) for the additive as well as the dominance regimes.[38] These results give a mechanistic rationale for the above conjecture without invoking different types of gene regulatory architectures between fitness and nonfitness traits, the concept of dominance modifiers, or any other type of additional mechanism or principle beyond what we already know about the functioning of gene regulatory networks. For a given trait, natural selection can just work straight on the available allelic variation underlying the spread of offspring values about the parent value.

By measuring allele-specific mRNA and protein expression patterns of genes underlying comparable fitness traits in few- and many-progeny species, and combining these with data on the coefficient of variation of offspring distributions in both groups, I would think we will have the experimental test bed needed to decide whether the above conjecture belongs to the class of embarrassing failures or not.

Heterosis and Regulatory Nets

According to Geiger,[78] classical and more recent analyses of generation means in several animal and crop species clearly demonstrate that genetic dominance in some form is by far the most important component of heterosis. The two earliest hypotheses regarding heterosis, the dominance hypothesis[79] and the overdominance hypothesis,[80,81] both based on single-locus theory, have competed for most of the last century. One would think that the introduction of allozyme markers, restriction fragment length polymorphisms, and high-density molecular linkage maps, should make it a relatively simple task to settle this debate. However, this has apparently not happened. Stuber et al[82] observed that heterozygotes for almost all the quantitative trait loci (QTL) for yield in maize had higher phenotypic values than the respective homozygotes. On the basis of data from an interspecific cross of rice, Xiao et al[83] suggested that dominance might be the basis of heterosis in rice. The importance of genetic dominance as the most important component of heterosis was challenged by Yu et al.[84] They investigated the genetic basis of heterosis in an elite rice hybrid by using a molecular linkage map with 150 segregating loci covering the entire rice genome. Overdominance was observed for most of the QTL for yield and also for a few QTL for the component traits. However, correlations between marker heterozygosity and trait expression were low, indicating that the overall heterozygosity made little contribution to heterosis. On the other hand, digenic interactions were frequent and widespread, and the results provide evidence that epistasis plays a major role as the genetic basis of heterosis. A connection between epistasis and dominance has also been reported in maize. De Vienne et al[85] found that epistatic interactions were involved in the control of 14% of the proteins investigated in maize coleoptile extracts. Doebley et al[86] found that two QTL with large effects on the aspects of plant and inflorescence architecture that distinguish maize and teosinte (being probably the loci for teosinte branched and terminal ear1 plus tassel replace upper-ear1, respectively) showed that maize alleles behaved in a more dominant fashion in maize background relative to teosinte background and that these QTL interact epistatically.

I think it is no coincidence that the model framework and the results by Omholt et al[38] seem to provide a platform by which the dominance, overdominance, and epistasis hypotheses may be reconciled to some degree. A one-locus negative autoregulatory structure is capable of generating dominance, and a positive autoregulatory one is capable of generating dominance as well as overdominance. On the other hand, through the notions of feedback-mediated epistatic genetic dominance and overdominance and down-stream mediated epistatic genetic dominance and overdominance it is clear that in many-locus regulatory networks the concepts of dominance and overdominance are intimately connected to the concept of epistasis. Thus heterosis may also be due to dominance or overdominance mediated by epistatic interactions realized within the same regulatory structure.

When two lines are inbred, they are likely to end up with a high degree of homozygosity at several regulatory loci with different types of alleles at some of these loci. The first generation hybrid line will be heterozygous for all these loci. From the above considerations it follows that the way the heterosis effect is realized is dependent upon the selection history and the allelic variation available before the selection started. That is, the same regulatory structure may generate heterosis that can be attributed either to dominance, overdominance, feedback-mediated epistatic genetic dominance and overdominance and down-stream mediated epistatic genetic dominance and overdominance, or other types of epistatic interactions. This possibility has to be investigated much more in detail by numerical experiments based on empirically confirmed network structures embedded in individuals exposed to selection regimes known to generate heterosis. The results from this theoretical exercise are likely to generate several predictions that can be tested in the wet lab.

Concluding Remarks

In this century much of biological research will become almost synonymous with the efforts to understand the functional expression of genes within the context of integrated biological systems. Biology is finally in a position to start revealing the causal links between genotype and phenotype in the wide sense. I expect that out of these efforts a theoretical structure that deserves to be called quantitative genetics will emerge. Its capabilities will include formalisms for detecting, describing, analysing and interpreting complex genetic systems underlying metric or meristic characters.

However, to achieve this, biological research will have to address phenomena at several systemic levels simultaneously, which will force it to become much more interdisciplinary and theoretically inclined. In this process its statistical, mathematical and computational tools will become substantially more sophisticated, and conceptual and methodological apparatuses, which are today almost totally separated, will become much more integrated.

The search for principles and methodologies that link the behaviours of molecules (i.e., genes) to system characteristics and functions (i.e., phenotypes) has been the prime occupation of genetics for the last 100 years. Pearson wrote to Galton in 1897: "...the solutions to these problems [variation, heredity, selection, and other phenomena relating to evolution] are in the first place statistical, in the second place statistical, and only in the third place biological" (quoted in ref. 87). I think it is fair to say that this does not give a representative description of the current situation in genetics. However, even though the move from explanatory models based on a beanbag genetics paradigm to models based on a paradigm acknowledging how genes actually work and interact represents a major change of focus methodologically and conceptually, it makes no change of disciplinary goal. Despite the presence of all the fancy words meant to describe various aspects of the current transition, we are still geneticists. Let us not forget that.

Acknowledgements

I am grateful to Erik Plahte, Tormod Ådnøy, Ben Hayes, Jon Olav Vik, Arne Gjuvsland, Theo Meuwissen, Sigbjørn Lien and Barbara Eriksen for their help with the manuscript.

References

1. Falconer DS. Introduction to quantitative genetics. 3rd ed. Harlow: Longman Scientific & Technical 1989.
2. Lynch M, Walsh B. Genetics and analysis of quantitative traits. Sunderland: Sinauer Associates, Inc., 1998.
3. Devlin B, Daniels M, Roeder K. The heritability of IQ. Nature 1997; 388(6641):468-471.

4. Fisher RA. The correlation between relatives on the supposition of Mendelian inheritance. Trans R Soc Edinburgh 1918; 52:399-433.
5. Kempthorne O. Status of quantitative genetic theory. Paper presented at: International Conference on Quantitative Genetics. 1977.
6. Cockerham CC. An extension of the concept of partitioning hereditary variance for analysis of covariance among relatives when epistasis is present. Genetics 1954; 39:859-884.
7. Kempthorne O. The correlations between relatives in random mating populations. Cold Spring Harbor Symp. Quant Biol 1954; 20:60-78.
8. DeWinter W. The beanbag genetics controversy: Towards a synthesis of opposing views of natural selection. Biol Philos 1997; 12(2):149-184.
9. Kempthorne O. An overview of the field of quantitative genetics. Paper presented at: Second International Conference on Quantitative Genetics 1988.
10. Lewontin R. Annotation: The analysis of variance and the analysis of causes. A J Hum Genet 1974; 26:400-411.
11. Cheverud JM, Routman EJ. Epistasis and its contribution to genetic variance-components. Genetics 1995; 139(3):1455-1461.
12. Deboer IJM, Hoeschele I. Genetic Evaluation Methods for Populations with Dominance and Inbreeding. Theor Appl Genet 1993; 86(2-3):245-258.
13. Fuerst C, James JW, Sölkner J et al. Impact of dominance and epistasis on the genetic make-up of simulated populations under selection: A model development. J Anim Breed genet 1997; 114:163-175.
14. Misztal I. Estimation of variance components with large-scale dominance models. J Dairy Sci 1997; 80:965-974.
15. Sehested E, Mao IL. Responses to selection on genotypic or phenotypic values in the presence of genes with major effects. Theor Appl Genet 1992; 85(4):403-406.
16. Burger R, Gimelfarb A. Genetic variation maintained in multilocus models of additive quantitative traits under stabilizing selection. Genetics 1999; 152(2):807-820.
17. Bentsen H. Genetic effects of selection on polygenic traits with examples from Atlantic salmon, Salmo salar L. Aquaculture and Fisheries Management 1994; 25:89-102.
18. Lewin B. Genes VI. 6th ed. Oxford: Oxford University Press, 1997.
19. Thieffry D, Huerta AM, Perez-Rueda E et al. From specific gene regulation to genomic networks: A global analysis of transcriptional regulation in Escherichia coli. Bioessays 1998; 20(5):433-440.
20. Shen-Orr SS, Milo R, Mangan S et al. Network motifs in the transcriptional regulation network of Escherichia coli. Nature Genet 2002; 31(1):64-68.
21. Lee TI, Rinaldi NJ, Robert F et al. Transcriptional Regulatory Networks in Saccharomyces cerevisiae. Science 2002; 298(5594):799-804.
22. Zhao XY, Hung MC. Negative autoregulation of the neu gene is mediated by a novel enhancer. Mol Cell Biol 1992; 12(6):2739-2748.
23. Knolle PA, Uhrig A, Protzer U et al. Interleukin-10 expression is autoregulated at the transcriptional level in human and murine Kupffer cells. Hepatology 1998; 27(1):93-99.
24. Morel Y, Mermod N, Barouki R. An autoregulatory loop controlling CYP1A1 gene expression: Role of H2O2 and NFI. Mol Cell Biol 1999; 19(10):6825-6832.
25. Hirata H, Yoshiura S, Ohtsuka T et al. Oscillatory expression of the bHLH factor Hes1 regulated by a negative feedback loop. Science 2002; 298(5594):840-843.
26. Tebarth B, Doedt T, Krishnamurthy S et al. Adaptation of the Efg1p morphogenetic pathway in Candida albicans by negative autoregulation and PKA-dependent repression of the EFG1 gene. J Mol Biol 2003; 329(5):949-962.
27. Arnone MI, Davidson EH. The hardwiring of development: Organization and function of genomic regulatory systems. Development 1997; 124(10):1851-1864.
28. Omholt SW, Plahte E. Feedback Confusions. J theor Biol 1994; 168(2):231-231.
29. Thomas R. Laws for the dynamics of regulatory networks. International Journal of Developmental Biology 1998; 42(3):479-485.
30. Plahte E, Mestl T, Omholt SW. Feedback loops, stability and multistationarity in dynamical systems. J Biol Syst 1995; 3:409-413.

31. Snoussi E. Necessary conditions for multistationarity and stable periodicity. J Biol Syst 1998; 6:3-9.
32. Gouzé JL. Positive and negative circuits in dynamical systems. J Biol Syst 1998; 6:11-15.
33. Glass L. Combinatorial and topological methods in nonlinear chemical kinetics. J Chem Phys 1975; 63:1325-1335.
34. Glass L. Classification of biological networks by their qualitative dynamics. J Theor Biol 1975; 54:85-107.
35. Snoussi E, Thomas R. Logical identification of all steady-states - the concept of feedback loop characteristic states. Bull Math Biol 1993; 55(5):973-991.
36. Mestl T, Plahte E, Omholt SW. A mathematical framework for describing and analyzing gene regulatory networks. J theor Biol 1995; 176(2):291-300.
37. Plahte E, Mestl T, Omholt SW. A methodological basis for description and analysis of systems With complex switch-like interactions. Journal of Mathematical Biology 1998; 36(4):321-348.
38. Omholt SW, Plahte E, Oyehaug L et al. Gene regulatory networks generating the phenomena of additivity, dominance and epistasis. Genetics 2000; 155(2):969-980.
39. Hill AV. The possible effect of the aggregation of the molecules of hemoglobin. J Physiol 1910; 40:IV-VIII.
40. Isaacs FJ, Hasty J, Cantor CR et al. Prediction and measurement of an autoregulatory genetic module. Proc Natl Acad Sci USA 2003; 100(13):7714-7719.
41. Damerval C, Devienne D. Quantification of dominance for proteins pleiotropically affected by opaque-2 in maize. Heredity 1993; 70:38-51.
42. Freeman M. Feedback control of intercellular signalling in development. Nature 2000; 408(6810):313-319.
43. Phillips PC. The language of gene interaction. Genetics 1998; 149(3):1167-1171.
44. Wolf JB, Brodie III ED, Wade MJ. Epistasis and the evolutionary process. Oxford: Oxford University Press, 2000.
45. Leonardi A, Damerval C, Devienne D. Organ-specific variability and inheritance of maize proteins revealed by two-dimensional electrophoresis. Genetical Research 1988; 52(2):97-103.
46. Wilkie AOM. The molecular basis of genetic dominance. J Med Genet 1994; 31(2):89-98.
47. Long AD, Mullaney SL, Reid LA et al. High-resolution mapping of genetic-factors affecting abdominal bristle number in Drosophila-melanogaster. Genetics 1995; 139(3):1273-1291.
48. Prioul JL, Quarrie S, Causse M et al. Dissecting complex physiological functions through the use of molecular quantitative genetics. J Exp Bot 1997; 48(311):1151-1163.
49. Frary A, Nesbitt TC, Grandillo S et al. fw2.2: A quantitative trait locus key to the evolution of tomato fruit size. Science 2000; 289(5476):85-88.
50. Gleason JM, Nuzhdin SV, Ritchie MG. Quantitative trait loci affecting a courtship signal in Drosophila melanogaster. Heredity 2002; 89:1-6.
51. Kopp A, Graze RM, Xu SZ et al. Quantitative trait loci responsible for variation in sexually dimorphic traits in Drosophila melanogaster. Genetics 2003; 163(2):771-787.
52. Pillen K, Zacharias A, Leon J. Advanced backcross QTL analysis in barley (Hordeum vulgare L.). Theor Appl Genet 2003; 107(2):340-352.
53. Thoday JM, Thompson JN. The number of segregating genes implied by continuous variation. Genetica 1976; 46:335-344.
54. Felsenstein J. Effect of linkage on directional selection. Genetics 1965; 52(2P1):349-363.
55. Lande R. Quantitative genetics and evolutionary theory. Paper presented at: Second International Conference on Quantitative Genetics 1988.
56. Turelli M. Phenotypic evolution, constant covariances, and the maintenance of additive variance. Evolution 1988; 42(6):1342-1347.
57. Gillespie JH, Turelli M. Genotype-environment interactions and the maintenance of polygenic variation. Genetics 1989; 121(1):129-138.
58. Rutherford SL, Lindquist S. Hsp90 as a capacitor for morphological evolution. Nature 1998; 396(6709):336-342.
59. Hartman JL, Garvik B, Hartwell L. Cell biology—Principles for the buffering of genetic variation. Science 2001; 291(5506):1001-1004.

60. Kitami T, Nadeau JH. Biochemical networking contributes more to genetic buffering in human and mouse metabolic pathways than does gene duplication (vol 32, pg 191, 2002). Nature Genet 2002; 32(4):681-681.
61. Veitia RA. A sigmoidal transcriptional response: Cooperativity, synergy and dosage effects. Biol. Rev 2003; 78(1):149-170.
62. Fisher RA. The possible modification of the response of the wild type to recurrent mutations. Am Nat 1928; 62:115-126.
63. Fisher RA. Two further notes on the origin of dominance. Am Nat 1928; 62:571-574.
64. Fisher RA. The evolution of dominance: Reply to professor Sewall Wright. Am Nat 1929; 63:553-556.
65. Fisher RA. The evolution of dominance. Biol Rev 1931; 6:345-368.
66. Fisher RA. Professor Wright on the theory of dominance. Am Nat 1934; 68:370-374.
67. Wright S. Fisher's theory of dominance. Am Nat 1929; 63:274-279.
68. Wright S. The evolution of dominance. Comment on Dr. Fisher's reply. Am Nat 1929; 63:556-561.
69. Wright S. Physiological and evolutionary theories of dominance. Am Nat 1934; 63:24-53.
70. Charlesworth B. Evidence against Fishers Theory of Dominance. Nature 1979; 278(5707):848-849.
71. Kacser H, Burns JA. The molecular-basis of dominance. Genetics 1981; 97(3-4):639-666.
72. Orr HA. A Test of Fisher Theory of Dominance. Proc Natl Acad Sci USA 1991; 88(24):11413-11415.
73. Grossniklaus U, Madhusudhan MS, Nanjundiah V. Nonlinear enzyme kinetics can lead to high metabolic flux control coefficients: Implications for the evolution of dominance. J theor Biol 1996; 182(3):299-302.
74. Porteous JW. Dominance—One hundred and fifteen years after Mendel's paper. J theor Biol 1996; 182(3):223-232.
75. Mayo O, Burger R. The evolution of dominance: A theory whose time has passed? Biol. Rev Cambridge Philosophic Soc 1997; 72(1):97-110.
76. Crnokrak P, Roff DA. Dominance variance: Associations with selection and fitness. Heredity 1995; 75:530-540.
77. Merila J, Sheldon BC. Genetic architecture of fitness and nonfitness traits: Empirical patterns and development of ideas. Heredity 1999; 83:103-109.
78. Geiger HH. Epistasis and heterosis. Paper presented at: Second International Conference on Quantitative Genetics 1988.
79. Davenport CB. Degeneration, albinism and inbreeding. Science 1908; 28:454-455.
80. East EM. Inbreeding in corn. Report of the Connecticut Agricultural Experimental Station 1907-1908. 1908:419-428.
81. Shull GH. The composition of a field of maize. Am Breed Assoc 1908; 4:296-301.
82. Stuber CW, Lincoln SE, Wolff DW et al. Identification of genetic-factors contributing to heterosis in a hybrid from 2 elite maize inbred lines using molecular markers. Genetics 1992; 132(3):823-839.
83. Xiao JH, Li JM, Yuan LP et al. Dominance is the major genetic-basis of heterosis in rice as revealed by QTL analysis using molecular markers. Genetics 1995; 140(2):745-754.
84. Yu SB, Li JX, Tan YF et al. Importance of epistasis as the genetic basis of heterosis in an elite rice hybrid. Proc Natl Acad Sci USA 1997; 94(17):9226-9231.
85. Devienne D, Maurice A, Josse JM et al.Mapping factors controlling genetic expression. Cell Mol Biol 1994; 40(1):29-39.
86. Doebley J, Stec A, Gustus C. teosinte branched1 and the origin of maize - evidence for epistasis and the evolution of dominance. Genetics 1995; 141(1):333-346.
87. Provine WB. The origin of theoretical population genetics. Chicago: Univ. Chicago Press, 1971.

Index

A

Adaptive selection 89, 98
Adh 91, 114, 115
Albright hereditary osteodystrophy (AHO) 25
Amorph 18
Amyloid A 28
Amyloidose 28
Aneuploids 89, 92, 96-98, 113, 115
Aneuploid syndrome 36, 89, 96-98
Arabidopsis 31, 98, 107
Autoregulatory locus 138, 139
Autosomal hypodontia 57

B

β-galactosidase reporter gene 53
Beta 2-microglobulin 28
Beta globin 27, 53
Butterfly color pattern 61, 63, 70, 72

C

Campomelic dysplasia 57
Cerebral malformation 57
Chaperone 31, 37, 42, 48
Chimeric proteins 23, 28
Chlamydomonas reinhardtii 4, 5, 21
COL5A1 52
Color patterns 61-63, 66, 70-72
Cryptogenic cirrhosis 27

D

Delta globin 27, 53
Developmental mapping 37-44, 46-48
Developmental noise 37, 38, 42, 44, 45, 47
Developmental stability 37, 38, 42-45, 47, 48
Dominance 1-5, 7-14, 17-24, 26, 27, 29-32, 36-41, 43, 44, 47, 58, 61, 62, 74, 75, 95, 98, 118, 119, 129, 131-133, 136, 138-150, 153
Dominant negative effect 23, 26, 27, 41, 144
Dosage compensation 13, 52, 55, 89, 91-93, 108, 113-115
Dosage effect 30, 89-93, 95-99, 112-115
Drosophila 2, 3, 29, 31, 91-94, 96, 100, 101, 107, 113, 114
Drosophila melanogaster 31
Duplex kidneys 58

E

Ehler-Danlos syndrome 52
Enzyme 2, 5-12, 14, 19-22, 25, 26, 38, 39, 70-72, 93, 111, 119, 122-126, 128, 129
Epidermolysis bullosa simplex 27
Epidermolytic palmoplantar keratoderma 27
Epistasis 18, 21, 22, 29, 37, 41, 42, 44, 45, 93, 100, 132, 133, 140-145, 149
Escherichia coli 119, 123-126, 128, 129, 136
Evolution 1, 3, 4, 12-14, 32, 44, 48, 75, 82, 89, 98, 99-101, 106, 108, 109, 111, 112, 115, 124, 129, 150
Evolutionary 2, 3, 12, 13, 21, 22, 32, 44, 57, 99, 107, 109, 112-114, 124, 128, 131, 133, 134, 147, 148

F

Familial hypercholesterolemia (FHC) 23
Feedback loop 136-142, 146, 147
Fetal hemoglobin 27
Fibrinogen 27, 28
Fisher's model 3-5, 12, 13

G

Gamma globin 27, 53
Gelsolin 28
Gene clustering 109, 113, 115
Gene dosage 23, 26, 30, 46, 58, 91, 92, 95, 114, 118, 144
Gene interaction 32, 89, 100, 133
Gene regulation 57, 70, 89, 99, 124
Greig cephalopolysyndactyly syndrome 57

H

Haemophilus influenzae 124, 128
Haploinsufficiency (HI) 13, 23-27, 29, 30, 38, 39, 41, 46, 52, 55, 57, 58, 61, 69, 75-77, 80-82, 85-89, 93, 144
Haplolethality (HL) 24
Heterosis 100, 101, 149, 150
Hexokinase 9-12
Hexokinase D 9-12
Homeobox gene 28, 57, 107
Hopeful Monsters 100
Hsp90 31, 32, 37, 42, 48
Huntington disease 28
Hybrid incompatibility 100, 101
Hydrocephalus 58
Hypermorph 18, 74
Hypomorph 18, 74

I

Ichthyosis bullosa of Siemens 27
Ig light chain 28
Imprinting 24, 25, 29, 99
Inverse effect 91, 92, 113, 114

K

Keratin 27

L

Loss-of-function allele 14
Loss-of-function mutation 22, 23, 52, 57, 124, 125, 129
Loss-of-function phenotype 41
Low density lipoprotein (LDLR) 23
Lymphadema-distichiasis syndrome 58
Lysozyme 28

M

Meesmann's corneal dystrophy 27
Metabolic control analysis (MCA) 21, 22, 25
Metabolic control analysis 7, 119
Metabolic flux 5, 7-9, 119, 123, 124, 127, 129
Michaelis-Menten reactions 8, 10, 11, 14, 69, 139, 147
Monilethrix 27
Mutations 2-5, 10, 12-14, 17-19, 21-32, 38, 41, 46, 52, 57, 58, 76, 77, 80, 81, 84, 87, 89, 93, 98-101, 109, 111, 113, 118, 124, 125, 127-129, 142, 144, 147
Myo2p 30

N

Negative control coefficient 8
NF1 46
Nonalellic noncomplementation 29
Nonlinear system dynamics 131, 135, 136
Nonlinearity 14, 18, 24, 32, 38, 39, 47, 48, 62, 74, 81

O

Oculopharyngeal muscular dystrophy (OPMD) 28
Ossification defect 57
Osteogenesis imperfecta 27
Overdominance 13, 14, 23, 29, 39, 40, 44, 139, 142, 149, 150

P

PABP2 28
Pachyonychia congenita 27
PAI-2 53
Pleiotropy 1, 12, 99
Polar overdominance 23, 29
Polymorphism 3, 13, 22, 134, 138, 142-145, 149
Prealbumin 28
Prion protein 28
Proα1(V) 52
Pseudo-bistable system 67, 68

Q

Quantitative genetics 131-136, 144, 145, 148, 150
Quantitative trait 20, 44, 89, 96-98, 100, 145, 149
Quantitative trait loci (QTL) 44, 89, 97, 98, 101, 145, 149

R

Recessivity 1-5, 7-10, 12-14, 17, 20, 23, 29, 129
Regulatory biology 131, 135, 136, 138, 140
Regulatory hierarchies 89, 96, 98, 99, 101
Regulatory nets 146, 147, 149

S

Saccharomyces cerevisiae 14, 118
SaethreChotzen syndrome 57
Selfish DNA pathway 112
Steady state 6, 7, 9, 10, 21, 52, 78, 79, 82-85, 87, 120-124
Stochastic gene expression 24, 46, 52, 54, 55, 58, 61, 62, 66, 70, 72, 74-76, 85
Stoichiometric network analysis 119, 122

T

Tooth agenesis 57
Transthyretin 28

U

Underdominance 39, 40, 44

V

Variance component 132-134, 145, 148

W

Waardenburg-Shah syndrome 57
White sponge nevus 27

X

X chromosome 13, 92, 93
X-linked gene 13

For Product Safety Concerns and Information please contact our
EU representative GPSR@taylorandfrancis.com Taylor & Francis
Verlag GmbH, Kaufingerstraße 24, 80331 München, Germany